D1784216

Engineering Construction and Materials

This book fills a long-felt need by both teachers and students for a single, comprehensive volume covering the whole of the work for the Engineering Construction and Materials syllabus in the City and Guilds of London Institute Mechanical Engineering Technicians Course, Part II. It will also be a handy work of reference for Part III examinations, and of use for some aspects of Engineering Design in the new H.N.C. in Engineering.

The materials section has been written to give the student an appreciation of materials in modern design. Certain metallurgical terms and studies have been worked into the text to aid interpretation of manufacturers' catalogues. Non-ferrous metals are introduced as a logical progression of elements added to a base metal; while the plastics section attempts to bridge the gap between the chemist who produces them and the engineer who uses them.

Cutting tool design is discussed in the light of the relevant British Standards, and the different cutting tool materials are introduced as a logical, historical progression, with the advantages, disadvantages and uses clearly laid out. Gauge design is similarly related to applications and materials usage. The engineering drawing section illustrates principles with many worked examples and problems. In the design chapters there is a new approach, with a serious attempt to convey to the student what the designer is thinking, with numerous opportunities for original thought and discussion.

Each chapter ends with a selection of exercises and a collection of general exercises completes the book.

Frank Race is on the staff of Barnsley College of Technology. John Hoyland is on the staff of the Engineering Industry Training Board.

Engineering Construction and Materials

by J. HOYLAND and F. RACE

CASSELL · LONDON

CASSELL & COMPANY LTD
35 RED LION SQUARE, LONDON WC1R 4SJ
SYDNEY, AUCKLAND
TORONTO, JOHANNESBURG

First published November 1968
Second edition May 1972

I.S.B.N. 0 304 29066 7

Phototypeset by BAS Printers Limited, Wallop, Hampshire
Printed in Great Britain by Ebenezer Baylis and Son Limited,
The Trinity Press, Worcester and London

372

Preface

This book is intended to cover the syllabus of 'Engineering Construction and Materials' for Part II of the Mechanical Engineering Technicians Course (Workshop Technology) of the City and Guilds of London Institute.

Within the very broad outline of the syllabus, we have endeavoured to cover all the topics relevant to the course in relation to the industrial experience and industrial needs of the student. For this reason, a chapter on plastics has been included as we consider that such a topic should not be omitted in view of present-day engineering applications of plastics materials.

Two omissions are distinctly noticeable. An explanation of Clauses 18 and 19 of B.S.308:1964, Engineering Drawing Practice, dealing with geometrical positional tolerances, has purposely been left out as we feel that the Standard is self-explanatory. Secondly, the use of B.S. Data Sheet No 1 'Primary Selection of Fits' has been excluded since its use is best explained by the teacher during the completion of detailed working drawings.

The student should not expect to find in this book all the answers to the exercises. We would like to encourage students to read further into the many books available on engineering and to discuss the various problems that arise with each other and with their teacher. It has been our intention to produce a book which obviates the needless and lengthy process of note taking. We hope that this book will take the place of much of the usual students' notes and allow more class time for work on the drawing board, for discussion and for research.

We wish to thank the British Standards Institution, the City and Guilds of London Institute and Hall and Pickles, Ltd for permission to reproduce diagrams and questions.

John Hoyland and Frank Race

Table of Contents

CHAPTER 1

Cast Irons

Cast iron is a very important engineering material which has been overshadowed by the rapid development of alloy steels. Ordinary cast iron has come to mean simply grey cast iron, useful for its high fluidity in the production of intricate castings. Grey cast iron, however, is only one form of this versatile material; rapid strides have been made in the last thirty years to produce other cast irons which include the great advantage of grey cast iron, namely, fluidity, as well as the tensile strength of steel. The development of spheroidal graphite cast iron has achieved this aim.

Grey cast iron is essentially an alloy of iron and carbon, with the latter usually in excess of $2 \cdot 5 \%$, and it is the form which the carbon takes that decides the mechanical and physical properties of the resulting metal. The form of the carbon is also used broadly to classify the irons into three main types, with a fourth type to cover the alloy or high-duty irons.

1 Grey Cast Iron

This is the most widely used iron in general engineering work and it derives its name from the colour of the cross-section of the metal. The carbon in this type of iron appears as graphite flakes dispersed throughout the structure and is known as free carbon, i.e. it has not combined with the iron. These flakes fill cavities in the iron and may be interconnected, giving rise to a broken structure. The inherent brittleness of the metal is due to these flakes which also cause a discontinuous chip to be produced when the metal is machined. However, there is some compensation: the graphite gives the material its self-lubricating properties.

The graphite flakes are set in a matrix which can consist of:
(*a*) ferrite, which gives a weak, easily cast and machined iron;
(*b*) ferrite together with pearlite, which gives a slightly stronger iron;
(*c*) pearlite, which gives the strongest iron.

Silicon is also found in iron. It acts as a softening agent by encouraging cementite to break up into ferrite and graphite. This can occur because cementite is actually a compound of iron and carbon; it is also very hard and brittle, having a white appearance under a microscopic examination. Cementite always contains approximately 6·5% carbon, the remainder being ferrite which simply refers to grains of pure iron.

When cementite and ferrite occur together in alternate layers, the resulting dark patches are known as pearlite. The layer structure is highly desirable because of the resulting high strength properties it conveys to the material.

Some of the main qualities of grey cast iron are:
(*a*) ease of casting, enabling intricate internal and external shapes to be cast (this is the most important property of the material because the casting process is the quickest method of obtaining a component from the raw material stage, e.g. automobile cylinder heads);
(*b*) good resistance to sliding wear, e.g. piston rings;
(*c*) noise and vibration damping capacity, e.g. pump mountings, gear boxes;
(*d*) tensile strengths from 150–400 MPa, associated with high compressive strengths giving rise to great rigidity, e.g. machine tool beds, columns and supports;
(*e*) resistance to soil, water and atmospheric corrosion, e.g. gas and water mains;
(*f*) good machining properties where high production rates are required, e.g. gear box casings for automobiles.

Because of these qualities, cast iron is an important basic material in all branches of the engineering industry.

2 Malleable Cast Iron

The basic metal for this second type of cast iron is not the normal grey iron, but an iron having a white or mottled appearance in section, known as white cast iron. It is an engineering material in its own right and deserves mention before malleable irons are dealt with.

Its white appearance is obtained by preventing any free carbon (graphite) from forming; this can be achieved by controlling the composition so that it contains little or no silicon and up to 3.5% carbon. The carbon combines with the iron to form iron carbide (cementite), producing a hard, almost unmachineable metal. Castings of this type are used where extreme abrasive conditions are to be met such as in stone- and glass-crushing machinery.

A second method of producing white iron is by rapid cooling, which prevents graphite from forming and is a valuable aid to manufacture. It is often desired to produce a harder, more wear-resistant surface on grey iron castings and this can be achieved by placing metal plates, called chills, into the mould. The molten metal cools quickly on coming into contact with them, forming white iron. The hard, fine-grained structure of this chilled surface metal also leads to a better surface finish after machining, usually by grinding.

There are several forms of malleable cast iron, as follows:

(*a*) *Whiteheart malleable iron.* This is produced by packing the white iron castings in boxes together with haematite (iron ore). The boxes are heated to 900°C, held at this temperature for several days and finally allowed to cool slowly. The treatment breaks down the cementite, allowing the carbon to appear in the free form. In thin sections, the free carbon is almost completely removed by the oxidizing effect of the haematite, leaving a metal with the properties of mild steel. In thicker sections, only the surface is decarburized leaving the interior of the casting with the carbon dispersed, not in flake form, but as tighter, round-shaped particles known as *nodules*; this removes the brittleness caused by the flake graphite.

(*b*) *Blackheart malleable iron.* In this process no decarburization takes place. The white iron castings are heated to 800°C in a neutral packing so that when the carbon is freed it becomes dispersed throughout the structure in a nodular form.

(*c*) *Pearlitic malleable iron.* This fairly recent extension of the blackheart process produces a malleable iron of high tensile strength which can be hardened and tempered. These properties are obtained either by the addition of up to 1% manganese to the initial composition, or by rapid cooling after the blackheart treatment. The manganese addition has the effect of stabilizing the iron carbide of the white cast iron so that some of it is retained to give a pearlitic matrix instead of the usual ferritic structure. However, some carbon is allowed to disperse, giving the nodular formation.

3

The rapid cooling technique ensures that some of the free carbon is retained as iron carbide to form the pearlitic structure. The quenching is not severe enough to alter all the carbon, so that nodules of graphite are still dispersed throughout a pearlitic matrix.

The material can be induction- or flame-hardened to give the surface a hardness of 700 Vickers Pyramid Number. This is appropriate for gears, pawls and cams.

Pearlitic malleable iron is suitable for any application requiring rigidity, tensile strength and shock-loading properties.

To sum up the malleable irons, the graphite present in the material is used initially to give fluidity to the molten metal during casting, and it is then dispersed by heat treatment into a form which does not produce brittleness. Qualities obtainable with malleable irons are:

    (i) increase in tensile strength: strengths in the region of 250–750 MPa are obtainable, for use in material handling equipment, e.g. conveyor chains;

   (ii) high ductility and shock resistance, e.g. for use in pipe fittings;

  (iii) good 'castability';

  (iv) some grades can be surface hardened.

These last two qualities are utilized in the manufacture of gears and sprocket wheels.

3 Spheroidal Graphite Cast Iron

The name of this third type of cast iron has not yet been standardized. It can be referred to as s.g. iron, nodular iron or ductile iron. However, since it is the graphite which is in the spheroidal form it would seem that s.g. iron is the most appropriate.

The manufacturing process has been developed since 1948 and produces the graphite in a nodular form by treatment of the metal in the ladle prior to casting. Unlike the malleable irons, the prolonged heat treatments involved in their manufacture are eliminated by chemical means during the casting stage of s.g. iron. The malleablizing process is limited to a maximum of 50 mm sections whereas there is no limit to the size of s.g. iron castings. Therefore heavy machine parts can be cast with all the advantages that the nodular form of graphite brings to the metal. It can be appreciated that there is bound to be some competition and overlapping of the two processes but in general malleable irons should be used for thin sections whilst s.g. irons are used for heavier sections.

The spheroidizing process consists of the addition of magnesium to the molten metal in the ladle. The magnesium has to be added in the form of a nickel alloy because unalloyed magnesium would produce a fierce reaction causing the molten metal to be splashed about, vapourizing the magnesium. The nickel-magnesium alloy contains 10–20% magnesium and the cast iron contains 1–2% of this alloy. Most of the commercial production of s.g. iron is based on the magnesium process.

The spheroidizing effect can also be obtained by adding small amounts of cerium to the ladle. In this method, however, the sulphur content of the cast iron must be carefully controlled, therefore the magnesium process is preferred.

S.g. iron is obtainable in different grades which are governed by the type of matrix structures obtained by various treatments:

(*a*) *Pearlitic s.g. iron.* This is the normal 'as cast' condition of the matrix in sections up to 50 mm. The structure consists of layers of ferrite and cementite with spheroids of graphite, for application where an iron of high tensile strength is required whose ductility and wear resistance are of less importance; e.g. for hydraulic cylinders and blast furnace charger bells.

(*b*) *Normalized pearlitic s.g. iron.* The mechanical properties of 'as cast' s.g. iron are considerably improved by heating to 800°C and cooling in air. Typical applications are hoist drums and clutch plates where a higher wear resistance is required.

(*c*) *Ferritic s.g. iron.* When the pearlite matrix is annealed the combined carbon breaks down and becomes distributed as spheroidal graphite in a ferrite matrix. In this state the iron is at its maximum toughness and ductility. The annealing is carried out by heating to 850–900°C for 1–4 hours followed by slow cooling to 650°C in the furnace before withdrawing. Typical applications are furnace doors, vice frames and jaws.

(*d*) *Hardened and tempered s.g. iron.* By quenching from 850°C in warm oil and tempering in the range 200–600°C, the properties of tensile strength, hardness and resistance to impact can be varied to suit desired purposes. Also s.g. iron responds readily to flame- or induction-hardening making it eminently suitable for gears and cams.

In summing up this particular cast iron, it can be said that it is a modern engineering material combining the qualities of excellent castability and machinability with high tensile strength, toughness and ductility.

In many applications it is replacing materials such as grey and malleable cast iron, non-ferrous alloys, cast and forged steel.

4 Alloy Cast Iron

There are many industrial applications in which the three types of cast iron so far discussed are not particularly suitable. Conditions of severe abrasion, corrosion and extreme temperature demand special properties. These can be obtained by the addition of various alloying elements such as nickel, chromium, molybdenum and copper.

(*a*) *Acicular irons.* By the addition of nickel ($1 \cdot 5$–$2 \cdot 0 \%$) and molybdenum ($0 \cdot 3$–$0 \cdot 6 \%$), the matrix of the cast iron can have an acicular or needle-like structure instead of the usual pearlitic one. This produces a high duty cast iron characterized by heavy sections and good wear resistance. Typical applications are die blocks for pressing or forging, and rolling-mill rolls.

(*b*) *Inoculated irons.* The late addition of calcium silicide to an iron which would normally cast 'white' produces a matrix approaching a $0 \cdot 9 \%$ carbon steel in strength and toughness. *Meehanite* is the trade name for such an iron. It possesses high strength with excellent machining properties, but the structure can be varied to meet different requirements, e.g. heat, wear and corrosion resistance.

Many uses can be found for the inoculated irons including hydraulic rams and cylinders, connecting rods, brake drums and glass forming moulds.

(*c*) *Austenitic Cast Iron.* This group of cast irons has the matrix rendered austenitic at room temperature by the addition of copper and nickel. These two elements lower the temperature at which austenite reverts back to cementite and pearlite to such an extent that austenite can exist even at normal room temperature.

The outstanding characteristics of these irons are:

corrosion resistance,
high temperature resistance,
non-magnetic and
low coefficient of expansion.

Austenitic cast iron can be obtained with the free carbon as flake graphite or in the spheroidal form, the latter combining the ductility and shock resistance of the s.g. irons with properties obtained from the austenitic matrix. Typical austenitic irons obtainable commercially are:

(i) *Ni-resist:* this is a trade name describing a non-magnetic austenitic grey iron with a composition including 14% nickel, 6% copper and 2% chromium. It is particularly useful for applications where a combination of heat and corrosion resistance is required, e.g. for valves used in chemical plants.

(ii) *Nicrosilal:* with an increase in the silicon content, the resistance to scaling at elevated temperatures is very much improved. This particular austenitic cast iron has a composition including 18% nickel, 6% silicon, and 2–4% chromium. It is suitable for temperature conditions up to 950°C. Obvious applications are furnace and boiler parts.

(*d*) *Martensitic cast iron.* The structure of a white cast iron consists of iron carbide and pearlite. The former being extremely hard whilst the latter is comparatively soft. To produce a cast iron with extreme wear resistance properties, it is necessary to harden the pearlite portion of a white cast iron. By the addition of suitable alloying elements, the hard constituent martensite can be substituted for the pearlite. Such a martensitic cast iron is *Ni-hard*, containing 3–4% nickel, 0·75–1·5% chromium, 3·5% carbon. It has been developed for parts where the casting process provides the necessary accuracy, because *Ni-hard* is virtually unmachinable. The applications of this iron are numerous, e.g. all crushing equipment, sand-blasting machine parts, in fact any application requiring extreme abrasion resistance.

Summary

From the range of cast irons mentioned, it can be seen that they present the engineer with a wide choice of properties for any application. They all have the common factor in that they are relatively easy to cast; after that it is a matter of presenting the carbon in the form necessary to provide whatever properties are required.

The following British Standard Specifications are available for cast irons:

Type of cast iron	B.S.S. number
grey cast iron	1452 : 1961
white cast iron	none
whiteheart malleable	309 : 1958

blackheart malleable	310 : 1958
pearlitic malleable	3333 : 1961
s.g. iron	2789 : 1961
austenitic cast iron	3468 : 1962
martensitic cast iron	none

EXERCISES

1. Describe the differences between the following forms of cast iron: grey (machining quality), whiteheart malleable, blackheart malleable and nodular (spheroidal graphite), from the following viewpoints: fluidity, compressive and tensile strength, impact resistance, carbon formation, heat treatment and machinability.

(C.G.L.I.)

2. Explain briefly the difference between malleable and spheroidal graphite cast iron.
   Choose an outline for which either material might be suitable and outline the factors which would be considered in making a decision which to use.

(C.G.L.I.)

3. Discuss the use of cast iron in the manufacture of machine tool beds. State clearly its particular advantages and disadvantages.

4. Give two examples of typical cast irons which contain one or more of the following alloying elements: nickel, chromium, molybdenum, and copper.
   For each of the cast irons, outline the properties obtained and give a typical application.

5. List the advantages which the use of iron castings offers to the designer in considering the choice of material for a particular application.

CHAPTER 2

Alloy Steels

Steel in its simplest form is an alloy of iron and carbon together with small amounts (trace elements) of silicon, manganese, sulphur and phosphorus.

In plain carbon steels the amounts can be as follows: silicon 0·5%, manganese 0·5–1·5%, sulphur and phosphorus 0·05%.

When these elements are in greater amounts, or there are purposeful additions of any other element, an alloy steel is produced.

The elements are added to improve certain properties of the steel:

(*a*) tensile strength and hardness at room temperature and elevated temperatures;

(*b*) resistance to shock, fatigue and abrasion;

(*c*) resistance to corrosion;

(*d*) ease of heat treatment, particularly in large sections.

Before the effects of some alloying elements are considered, it is important to understand two metallurgical terms relating to alloys: *eutectic* and *eutectoid*.

These words are derived from the Greek—*eutektikos* meaning 'easily melted'. Basically, a eutectic alloy is one in which metals are soluble in the liquid state but insoluble in the solid. Thus atoms which have freely intermingled whilst in liquid form seek out and combine with other atoms of their own kind to form separate groups or crystals on solidification. This produces a very fine layer structure of one metal on top of the other, and under microscopic examination each metal is readily identified. The melting point of this structure is below that of the individual metals concerned and is known as the eutectic temperature. An equilibrium diagram,

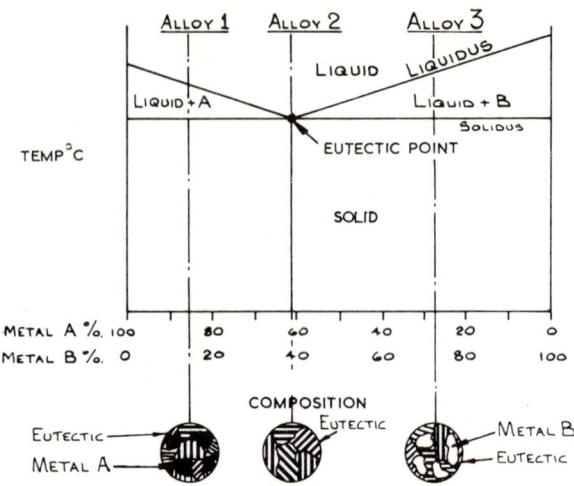

Fig. 1 Equilibrium diagram for two metals completely insoluble in the solid state

which is a graphical representation of the cooling characteristics of a system of alloys, shown in Fig. 1, has two metals A and B which are alloyed together and the sketches show micro-structures of three separate alloys taken from the diagram. In nos 1 and 3, the the major alloying metal is precipitated out until the eutectic temperature is reached, at which point the remaining liquid solidifies to give the familiar eutectic layer structure. A point to note is that alloy no. 2 is wholly eutectic and has no pasty stage during solidification, i.e. it solidifies instantly as it cools. In the lead-tin range of alloys, i.e. soft solders, a eutectic composition is obtained at 40% lead, 60% tin, and represents the lowest melting point in the system at 183°C.

When structural changes take place after solidification, i.e. during cooling to room temperature, and a layer-type structure is formed, the alloy is known as a eutectoid and the temperature at which this occurs is known as the eutectoid temperature. A eutectoid is obtained in a 0·87% carbon steel below a temperature of 720°C, when the structure is wholly pearlitic.

Therefore:

(*a*) a eutectic structure is produced on solidification;

(*b*) a eutectoid structure is produced after solidification.

Solid solution

A solid solution alloy consists of metals which are soluble in each other in both the liquid and solid state. Thus, on cooling, the different metals cannot be identified by microscopic examination as each grain shows the same characteristics with nothing to suggest the presence of different metals.

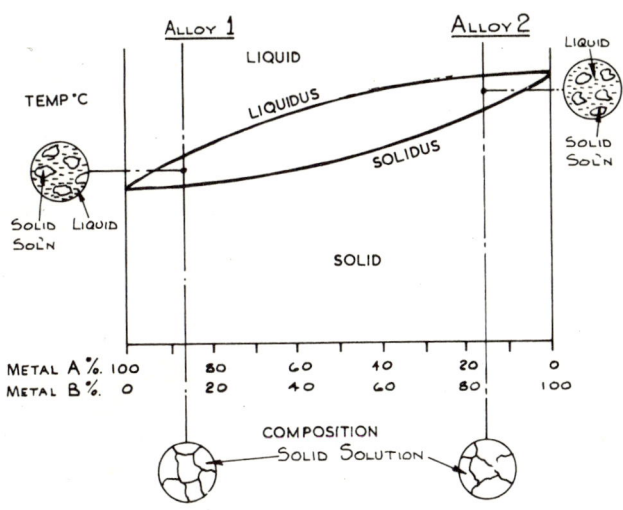

Fig. 2 Equilibrium diagram for two metals completely soluble in the solid state

Fig. 2 shows a typical equilibrium diagram for an alloy producing a solid solution. At any stage of the solidification of an alloy of this type, only crystals of the alloy are precipitated and never crystals of the original metals.

In the familiar steel portion of the iron-carbon equilibrium diagram, Fig. 3, the austenite is a solid solution of iron and carbon whilst the pearlite is a eutectoid of iron and iron carbide. The ferrite is crystals of pure iron and the cementite is a form of iron carbide.

As a general rule, solid solutions are more ductile and resist corrosion whereas the eutectic structure provides strength.

11

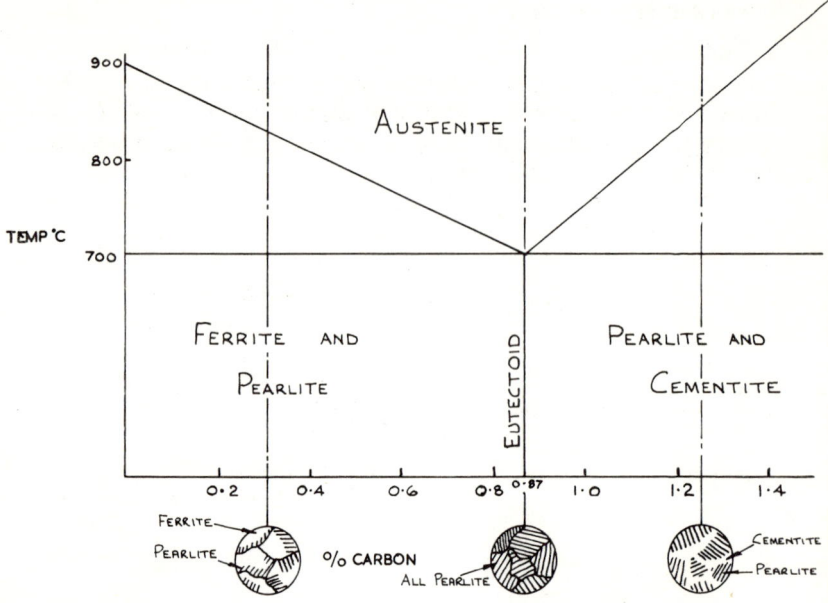

Fig. 3 Portion of the iron-carbon equilibrium diagram

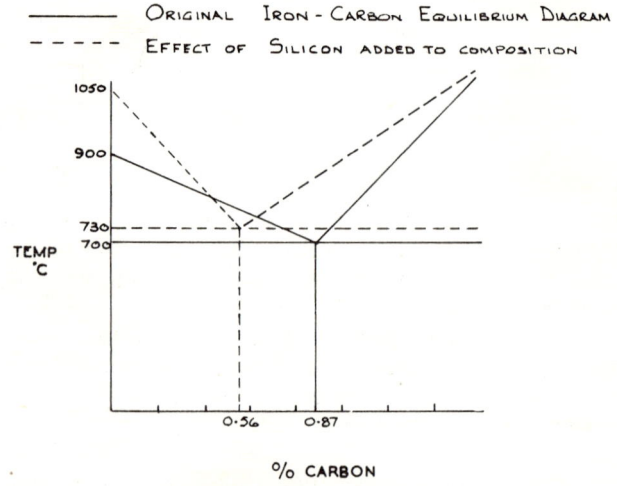

Fig. 4 Effects of 2% silicon

Effect of alloying elements

1. Silicon

This element is present up to 0·5% in all steels. It does not form a carbide but dissolves in the ferrite to form a solid solution. With amounts greater than 0·5% the following effects occur:
(a) the critical points in the iron-carbon diagram are elevated;
(b) the amount of carbon required for eutectoid composition is reduced (see Fig. 4);
(c) the silicon hardens and strengthens the ferrite and provides these benefits for the steel;
(d) it improves toughness and resistance to fatigue.

The three main types of silicon steel in use are:
(a) Spring steel: a typical composition is 0·3–0·6% carbon, 1·5–2·6% silicon, 0·7–1·0% manganese.
(b) Valve steel: 0·4% carbon, 3·5% silicon, 8% chromium. The resistance to oxidation at elevated temperatures is increased.
(c) Laminated cores for transformers and magnets: 0·05% carbon, 3·5% silicon. The magnetic and electrical losses are reduced.

The disadvantages of silicon as an alloying element in steel are that it coarsens the grain size and also tends to break down iron carbides into ferrite and graphite.

2. Manganese

This is present in all steels in amounts up to 0·5% as an impurity and up to 1% when added intentionally to act as a deoxidizer during the melting process. It has the additional benefit of reacting with any sulphur present to form manganese sulphide, which appears as grey spherical particles dispersed throughout the microstructure. Whilst this reduces the tensile strength, it produces excellent machining properties, giving rise to a discontinuous chip formation. This is one type of free-cutting steel and is listed as En 1B (B.S. 971) with a composition of 0·07–0·15% carbon, 1·0–1·4% manganese, 0·3–0·6% sulphur.

Manganese is considered to be an alloying element when in excess of 1·5%. With this amount there is sufficient for part to pass into solution with the ferrite and part to combine with the carbon to form manganese carbide, producing the following effects:
(a) the change points in the iron-carbon diagram are considerably depressed;

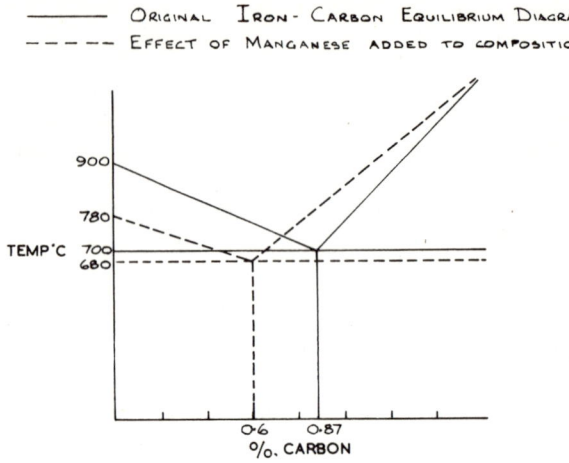

———— Original Iron - Carbon Equilibrium Diagram
— — — — Effect of Manganese added to composition

Fig. 5 Effects of 2·5% manganese

(*b*) the amount of carbon for eutectoid formation is reduced (see Fig. 5);

(*c*) by the formation of a solid solution with ferrite, the steel is considerably strengthened.

To produce a plain carbon steel in a hard condition, the metal is heated to some temperature where the carbon present is taken into solution with the iron, i.e. austenite is formed, and then a severe quenching is applied. The rate of cooling prevents the carbon from reverting back to cementite and a new structure is formed known as martensite, which appears as a dark needle-like formation. Due to it being a distorted formation it is no surprise that the hardness is accompanied by extreme brittleness. When the change points in the diagram are depressed, the rate of cooling from the austenitic condition to produce martensite is decreased.

The cooling rate is very important in the heat treatment of all steels and is known as the critical cooling velocity. With a reduced cooling rate, less severe quenching media can be used, e.g. oil or air, which minimize the risk of distortion or cracking.

The general effect of the manganese is therefore to produce steels capable of being hardened with relatively slow rates of cooling, with a tendency, also, for the metal to retain dimensional accuracy after hardening and tempering. This is a useful property when intricate dies have to be produced on which no further grinding or machining is possible after heat treatment. Such steels are termed non-shrinking steels, a typical composition for this type of steel being 1% carbon, 2% manganese.

The high-manganese steels with a content of 12–14% manganese can, when quenched from 1000°C, possess a tough austenitic structure at room temperature, i.e. the carbon remains in solution with the iron. If such a steel is subjected to abrasive or cutting action the effect is that the austenite at the working surface will decompose, producing martensite, and thereby increasing hardness and wear resistance under service conditions. The high-manganese steels are used for extreme wear conditions, as required by such items as stone-crusher jaws, railway points, lips of earth-moving buckets and armour plate.

Wherever possible, due to the obvious difficulty in machining them, the high-manganese steels are cast to shape. However, machining by drilling, turning, shaping and planing is possible under certain conditions. It is important to have a high temperature at the cutting edge because the work-hardening diminishes as the temperature increases. This means that no coolant can be used and it imposes limitations on the type of cutting-tool steel used. Generally, to withstand severe cutting temperatures, cobalt super-high-speed steel tool bits are used. These are ground with approximately 7° clearance, 12° true rake and a generous approach angle providing a long cutting edge. Cutting speeds in the region of 4–5 metres per minute are recommended, together with absolute rigidity of both machine and tool.

Any slow cooling of these steels should be avoided, since the precipitation of carbides at the grain boundaries causes embrittlement. Any heating involving temperatures in excess of 250°C should be followed by a severe water quench.

3. Nickel

This element is considered to be in an alloying amount when in excess of 0·5% and produces the following effects:

(*a*) it depresses the change points in the iron-carbon diagram;

(*b*) it reduces the amount of carbon for eutectoid formation;

(*c*) it does not form a carbide but passes into solid solution with the ferrite, therefore strengthening the resulting steel;

(*d*) it increases the resistance to oxidation at elevated temperatures.

Nickel steels are noted for their high strength, ductility and

15

toughness, a typical composition is En 22 which has 3·5% nickel, 0·4% carbon, providing a strong, shock-resisting steel. This is used to advantage in providing a case-hardening steel such as En 33 with a composition of 0·1% carbon, 2–3% nickel. The presence of the nickel restricts the grain growth by the prolonged heating during the carburizing stage of the process, thus eliminating a core-refining treatment.

The very high percentage nickel steels are, of course, expensive, but find many applications in industry. One such steel has a nickel content of 36% and is known as Invar steel. The nickel effects the coefficient of expansion to such an extent that it is almost zero between 0°C and 100°C. The steel finds application in sensitive instruments such as clocks and measuring devices.

One disadvantage of using nickel as an alloying element is that it has a tendency to break down any iron carbides into iron and graphite; thus considerably weakening the structure. For this reason the carbon content for a nickel steel is kept under 0·4%. Where this is not possible then another element must be added to prevent graphitization; chromium is such an element and finds many uses as the ideal partner to nickel in the manufacture of alloy steels.

4. Chromium

This element is considered to be in an alloying amount when in excess of 0·5% and can be added in amounts up to 35%. Steels containing more than 11% of this element tend to form an inert layer on the surface which resists attack by oxidation; hence its important use in the manufacture of stainless steel which will be shown in a later section.

The effects of chromium as a simple alloying element are as follows:

(*a*) it dissolves in the ferrite, promoting increased hardness and strength;

(*b*) it combines with carbon to form complex carbides known as *double carbides* which are hard and extremely wear resistant;

(*c*) it reduces the amount of carbon for eutectoid;

(*d*) it raises the change points in the equilibrium diagram.

Typical application of carbon-chromium steels are: En 31: a 1% carbon, 1·3% chromium steel used for ball bearings (this steel requires an initial heat treatment before hardening to break down the massive carbides formed); and En 11: a 0·6% carbon, 0·6% chromium steel used for heavy duty gears.

A disadvantage in the use of chromium is that grain growth, i.e. when grains combine to form larger grains, is extremely rapid after the austenitic range has been reached, therefore the heat treatment temperatures must be very closely controlled.

Nickel-chromium steels

Nickel steels are noted for their high strength, ductility and toughness whilst chromium steels have hardness and resistance to wear. The combination of the two elements produces a range of steels possessing all these properties. They provide the engineer and metallurgist with an almost ideal partnership in that the features of one offset the disabilities of the other. For example, the rapid grain growth associated with chromium is countered by the inhibiting effect of the nickel, also the graphitizing effect of nickel is balanced by the carbide-forming effect of chromium.

There is one factor which prevents this becoming a perfect relationship and that is a serious defect known as *temper brittleness*. When a nickel-chromium steel is held or tempered between 250°C and 400°C, a pronounced drop in impact value is experienced. A tough condition is regained by heating to above 600°C and rapidly cooling, though generally the addition of 0·5% molybdenum reduces the tendency to temper brittleness.

A common steel in this combination of alloying elements is En 24: 0·4% carbon, 1·3–1·8% nickel, 0·9–1·4% chromium, 0·3% molybdenum. This provides a good, tough steel which is used for crankshafts, connecting-rods and other high duty applications.

Other steels in this range are En 30A: $4\frac{1}{4}$% nickel, 1·3% chromium, 0·3% carbon, and En 30B which is basically the same with the addition of 0·3% molybdenum. These steels are recommended where the minimum distortion is required during heat treatment.

Most steels in the nickel-chromium range are used in the hardened and tempered condition.

Stainless steels

These are corrosion resisting steels where the chromium content has been increased to above 11% to provide a surface layer of oxides which are impervious to most corrosive media. This oxide layer is colourless and, unlike most oxides related to steel, it is very tenacious.

There are three main types of stainless steel and they can be classified and named according to their structures.

(*a*) *martensitic:* this type contains 12–14% chromium with varying amounts of carbon depending upon the service conditions required. A carbon content of 0·2% is used for steam valves where hardness is not a vital factor but table cutlery and surgical instruments requiring hardness and a good cutting edge contain 0·3% carbon. Ball-bearings and springs working under corrosive conditions have up to 2% carbon.

(*b*) *ferritic:* this type is often referred to as a stainless iron because the structure is wholly ferritic with carbide particles. It cannot be hardened by heat treatment but optimum corrosion resistance is obtained by quenching from 875°C. The chromium content ranges from 16–30%.

The higher content is used where corrosion resistance takes priority over strength and toughness, whereas the lower chromium content is used for such items as sink units where ductility with some corrosion resistance, is required.

The carbon content varies from about 0·1% for the wrought chromium irons to 3% for cast chromium irons.

(*c*) *austenitic:* the most used alloy of this type is the 18 : 8 i.e. 18% chromium with 8% nickel. The nickel content suppresses the change points to such an extent that an austenitic condition appears as the steel is produced. To obtain maximum softness the steel is quenched from 1000°C producing a ductile material which can be cold worked. In this condition it work-hardens rapidly, but provides a general corrosion-resistant material useful in tubing and sheet form.

A difficulty in fabricating this type of stainless steel is the defect known as *weld decay.* When the steel is held in the temperature range 500–600°C, chromium carbides are precipitated out at the grain boundaries with the result that areas of lowered chromium content adjacent to the grain boundaries have their resistance to corrosive attack reduced; this is shown by a band of corrosion appearing near the edge of the weld. An increased carbon content tends to increase the susceptibility to weld decay therefore the carbon is usually held at 0·08%.

By reheating the affected parts to 1000°C and quenching, the carbides are taken back into solution and the corrosion resistance is restored. Where this heat treatment is not practical, elements such as titanium and columbium can be added to the alloy, having a greater affinity for carbon than the chromium.

5. Molybdenum

This element stabilizes the carbide formation preventing graphitiza-tion, therefore it is used in conjunction with elements which tend to break down the carbide structure. Where molybdenum is present in amounts up to 0·7% the properties of hardenability, toughness and strength of the steel are improved. The addition of molybdenum to a nickel-chrome steel renders that steel less prone to temper brittleness.

6. Tungsten

The main contribution of this element to alloy steels is the ability to induce a property known as *red hardness*. It means that the change points are raised to such an extent that properties such as hardness can be retained even at red heat producing a range of cutting tool steels known as high-speed steels. A typical composition is 20% tungsten, 4% chromium, 10% cobalt, 0·7% carbon.

The heat treatment of high-speed steel to obtain solution of the carbides of tungsten and iron is complex. It consists of slowly heating to 800°C, rapidly heating from 800°C to 1250°C followed by quenching in a dry air blast. A secondary hardening is required which consists of heating to 600°C and allowing to cool slowly.

In conclusion, it has to be stated that no section covering alloy steels can ever be complete. The steel industry is being constantly called upon to produce alloys which have to meet ever changing conditions of service. The construction of atomic powered ships and submarines, and the storage at sub-zero temperatures of industrial gases in a liquid form, are just two modern examples where a special challenge has had to be met.

In order to keep abreast of the development of special duty alloy steels, the reader is advised to consult the available literature circulated by the industry itself.

EXERCISES

1. Give two examples of typical steels which include one or more of the following alloying elements:
   tungsten, molybdenum, chromium and nickel. For each of the steels, outline the properties and give a typical engineering use.

(C.G.L.I.)

2. In the heat treatment of steel, explain what is meant by the following terms:
   (*a*) critical cooling rate;
   (*b*) mass effect;
   (*c*) limiting section.

3. Stainless steels can be classified into three groups. Give the composition and application of one steel in each group. The welding of one of these steels may present certain difficulties. Describe the nature of these difficulties and say how they may be overcome.

(C.G.L.I.)

4. Explain fully what is meant by a free-cutting steel. To obtain the machining property 'free-cutting' certain elements can be added to a steel. Describe the effect of any three of these elements and the types of steel in which they would be used.

5. Write brief notes on the following alloy steels: En 25 and En 57.

CHAPTER 3

Non-Ferrous Alloys

1 Copper and its alloys

The properties of copper are very much affected by its degree of purity. The metal is obtained by extraction from its ore, followed either by refining in a smelting furnace or by electrolysis. Electrolytic copper is the purest form and is used in the production of high-duty alloys and for high-performance electrical conductivity in wires and cables. By the addition of small amounts of other elements, special properties can be obtained and examples of each type of copper follow:

Cadmium copper: by the addition of up to 1% of cadmium the strength, toughness and fatigue properties are increased with only a very slight reduction of conductivity. It is used for long-span overhead transmission lines and electrical contacts.

Arsenical copper: contains 0·5% arsenic, which increases its tensile strength and maintains that strength for temperatures up to 300°C. The resistance to atmospheric corrosion is also increased, but with a decrease in conductivity. It is used in boiler tubes, fireboxes etc.

Beryllium copper: contains up to 2% beryllium which provides a copper which can be heat-treated so that a soft condition is obtained for working the alloy; a further heat treatment can produce hardness and strength. It is used for springs, flexible bellows and components for measuring devices.

However, the main use of copper is as an alloying element in conjunction with large amounts of one or more other elements. The two most important of these are brass, an alloy of copper with zinc, and bronze, an alloy of copper with tin.

Brasses

These alloys can be broadly classified into two groups, based upon the amount of zinc present; those containing up to 37% zinc are known as the 'alpha' (α) brasses.

The second group are those with more than 37% zinc. They are less ductile but are especially suited for hot-working processes, and are known as the 'alpha-beta' ($\alpha + \beta$) brasses.

(*a*) *Alpha* (α) *brass:* when zinc is alloyed to copper the result is a solid solution. If the zinc content is less than 37% the structure is the normal homogeneous type, typical of solid solutions. The metallurgists have given this the title of α-phase. It seems that the practice of naming structures after their discoverers did not spread from the ferrous to the non-ferrous alloys.

Since this type of brass is a perfect solid solution it responds very readily to cold-working processes, the most ductile alloy in the group being the 70 : 30 alloy known as *cartridge brass*. It is used for all deep-drawing or severe cold-deformation applications, producing accuracy of dimensions and good surface finish.

According to the amount of cold-working or reduction in cross section, it may be necessary to apply an inter-process treatment to relieve brittleness and surface stress. This consists of heating the metal to just below 280°C and cooling it either in water or air, the rate of cooling being unimportant. All highly stressed brass components must be subjected to this process prior to being put into service in corrosive industrial conditions in order to prevent *season-cracking* or *stress corrosion*. The corrosion attack aggravates any slight ruptures left by cold-working.

Cartridge brass is obtainable in different tempers or degrees of hardness. Soft temper is a fully annealed material, i.e. heated to between 300°C and 600°C and cooled in air. A small amount of cold-work produces quarter-hard and half-hard temper, followed by hard and spring hard tempers as the cold-work is increased.

Other alloys in this group are:

   (i) Basis brass: 36% zinc; which provides a cheap material for general presswork.

  (ii) Admiralty brass: 70% copper, 29% zinc, 1% tin; used for condenser tubes and applications requiring resistance to corrosion.

 (iii) Aluminium brass: 76% copper, 22% zinc, 2% aluminium; provides excellent resistance to corrosion in steam power-plant tubing etc.

(iv) Gilding brass: 80% copper, 20% zinc; used for decorative purposes and can be heavily cold-worked.

(b) *Alpha-beta* ($\alpha + \beta$) *brass:* with increase in the proportion of zinc in excess of 37%, a second solid solution appears, known as beta (β)-phase, resulting in a duplex structure of α and β. The beta crystals are hard and give an increase in tensile strength with a decrease in ductility. When the metal is heated, however, the beta constituent renders the brass very plastic so that it becomes readily workable by hot-rolling, extruding etc. The main alloy used in this group is known as *Muntz metal*, with a composition of 60% copper, 40% zinc, and finds application as hot rolled plate, forgings and castings. A variation of this is *naval brass* with approximately 1·5% tin for use in corrosive conditions.

The $\alpha + \beta$ group are difficult to machine but by the addition of 3% lead to these alloys a range of free-cutting brass is obtainable. The lead is insoluble in the brass and is distributed throughout the structure as minute globules, which produce a discontinuous chip formation on machining, with a good surface finish.

Additions of elements such as manganese, iron, aluminium and tin to the 60 : 40 group produce a range of alloys known as *high-tensile brasses*. Where hardness and strength are required these additions are usually in the range 0·5–3%. A typical composition is 56–60% copper, 1% tin, 2% manganese, 1% iron and 1·5% aluminium, which is used for forgings and hot worked sections. Variations in the elements improve casting properties whilst maintaining toughness and strength. Applications in the cast form are marine propellers, axle boxes, pump bodies and gear wheels, and in the wrought form, spindles, shafts, nuts and bolts. Very often the term *manganese bronze* is applied to this particular group of brasses but to avoid confusion it is better to refer to them as high-tensile brasses.

Bronze

A true bronze is an alloy of copper and tin together with small amounts of other elements such as lead, nickel and zinc. There is a trend towards referring to alloys of copper and elements other than tin, as bronzes. However, for the sake of clarity and conformity we will refer to the true bronzes as tin-bronzes. They are used where the properties of good casting, wear and corrosion resistance are important.

A tin content of up to 8% produces an α-solid solution which,

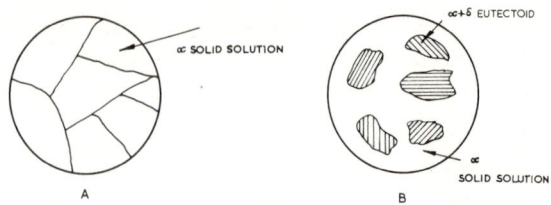

Fig. 6 Microstructures of bronzes. A. Tin content less than 8%. B. Tin content 8–15%

like the α-brass, is ductile and provides a range of wrought bronzes which can be cold-drawn and rolled, but in this respect they are inferior to the wrought brasses. To improve mechanical properties small amounts of phosphorus, up to 0·3%, are often added to produce a wrought phosphor bronze suitable for springs and clips.

An increase in the tin content beyond 8% introduces a secondary structure amongst the α-phase solid solution. This newcomer is a eutectoid of α with another phase referred to as delta (δ), see Fig. 6. The eutectoid formed is much harder than the solid solution in which it is dispersed, and provides the ideal requirements for a bearing material. The softer matrix wearing preferentially, provides channels for the distribution of lubricant, whilst the harder islands take the bearing load. The useful bearing alloys lie in the range 8–15% tin because above 15% the increasing amount of eutectoid promotes brittleness which reduces impact and fatigue properties.

The bearing bronzes are usually sand-cast to the required shape, although small bushes and bearings are often chill-cast as rods or tubes. The range of properties available in these bronzes is extended by the addition of the elements phosphorus, lead and zinc.

(*a*) *Phosphor-bronze:* the phosphorus is added for two purposes firstly as a de-oxidizer during the melting and casting process, to improve mechanical properties; the second purpose is that when it is in excess of the amount required for deoxidation, phosphorus reacts with copper to form a hard compound of copper phosphide. Under a microscope this compound can be seen dispersed in the eutectoid thus increasing the wear resistance. The phosphorus content is kept in the range 0·1 to 1%, due to casting difficulties and the fact that a large amount of copper phosphide present would induce shaft wear.

(*b*) *Leaded bronze:* lead is insoluble in copper-tin alloys and appears in the micro-structure as widely distributed small globules. Up to 5% lead is added to improve the machining qualities, but lead in excess of this amount reduces the dry coefficient of friction by providing a measure of self lubrication to the bearing. One of the most used alloys in this range has a composition 80% copper, 10% tin, 10% lead. Other bronzes are in use with a lead content varying up to 30% and are sometimes referred to as *plastic bronzes*. The high lead content reduces the strength and wear resistance and their application is chiefly as linings between a bearing and its shell.

(*c*) *Gun-metal:* this is an alloy of copper, tin and zinc which is widely used for intricate castings. The zinc renders the molten alloy more fluid and also acts as a deoxidizer by forming a zinc oxide scum on the surface of the melt. A typical alloy in this range is known as Admiralty gun-metal having a composition of 88% copper, 10% tin, 2% zinc, providing a standard alloy for marine castings, valve and pump bodies. Where the service conditions are less severe, a leaded gun-metal may prove adequate, with the advantages of reduced cost, improved castability and good machinability. Typical composition 85% copper, 5% tin, 5% zinc, 5% lead.

Other copper alloys

(*a*) *Aluminium bronze:* This type of bronze does not contain tin but is a straight copper-aluminium alloy, with the aluminium being of the order 2–12%. Once again a familiar pattern arises as the aluminium content is varied inside the range mentioned: 2–7% produces the α-phase solid solution which provides a cold-working alloy having high corrosion resistance. It is used in heat exchangers and condensers, etc.

Increase of the aluminium content provides a hot-working or casting alloy having the duplex structure of α-solid solution with α + δ eutectoid. This alloy has high resistance to corrosion and wear, coupled with high strength, and will maintain these properties at elevated temperatures. However, the difficulty experienced in the casting process in the early development of this alloy has probably built up a slight prejudice against its use. Needless to say, at the present time no such difficulties exist, and the designer has a strong heat resisting material available to him. Another property of the aluminium bronzes is that they can be heat-treated to produce a martensitic structure similar to that found in steels; therefore, by quenching from 900°C and using a tempering treatment, the toughness and hardness can be varied to suit the service conditions.

In addition to the standard aluminium bronzes there is a range of complex alloys available which contain additions of nickel, iron and manganese. A typical analysis is: 10% aluminium, up to 6% nickel, 4% iron and 2% manganese, this produces an alloy with a very high resistance to abrasion, wear and corrosion coupled with high tensile strength of the order 700 MPa. These alloys are finding increasing use both as sand and die-castings.

(*b*) *Cupro-nickel:* the alloying of the two metals copper and nickel produces a solid solution throughout the full range of composition, and provides a series of wrought alloys. An increase in the nickel content brings about an increase in the corrosion resistance of the material; up to 2% nickel provides an alloy suitable for boiler firebox stay rods. 12–30% nickel produces an alloy with excellent drawing properties which is used for such applications as condenser tubes and sea-water pipelines. Higher nickel contents, i.e. between 40–45%, produce alloys with high electrical resistance and resistance to oxidation at elevated temperatures, e.g. Constantan (60% copper: 40% nickel).

(*c*) *Nickel silvers:* these are alloys of copper, nickel and zinc in widely varying proportions, e.g. copper 55–65%, nickel 5–35%, zinc 10–35%. They contain no silver but are so named because of their silvery appearance. Where the zinc content is low, the alloys can be cold-worked and find wide application in cutlery and automobile, marine and decorative fittings. Where the zinc content is high the metal cannot be cold-worked and is used principally for decorative castings.

2 Aluminium and its alloys

Aluminium is available in various grades of purity, from 99% to 99·8% with 99·99% in special circumstances, and its properties can vary with the degree of purity. For the designer it offers many attractive features:

(*a*) low specific gravity: 2·68 as compared with 7·8 for steel;

(*b*) it promotes its own shield of oxide to resist corrosion by the atmosphere or under marine conditions. This shield can be improved artificially by an electrolytic process known as *anodizing*;

(*c*) excellent electrical conductivity, offering 60% that of copper.

Other lesser qualities are that it has good thermal conductivity, and is non magnetic and non-toxic; the latter quality is useful in food wrappings and container linings.

The principal disadvantage of commercially pure aluminium is its relatively low strength, i.e. in the soft condition it has a tensile strength of 6 tonf/in². However, by the addition of the various elements—copper, magnesium, zinc, silicon manganese—the strength of 90 MPa. However, by the addition of the various pairing the inherent advantages of the aluminium base.

For convenient consideration, aluminium alloys can be classified under two main groups, wrought and cast. Each of these is sub-divided into those which respond to heat-treatment and those which do not.

Wrought Alloys

(*a*) *Non-heat-treatable:* the strength of these alloys is obtained by work-hardening in such processes as rolling sheet, or drawing tubes and wire. The properties obtained depend upon the amount of cold-working applied and therefore a range of 'temper' is available: soft, quarter-hard, half-hard and hard.

The two main alloys available in this group are: $1-1\frac{1}{2}\%$ manganese, which greatly increases the strength whilst retaining cold working properties and is used for building and transport panels as well as in packaging; and $2-7\%$ magnesium, a tough, strong alloy with exceptional resistance to marine conditions, having a tensile strength varying from 180–370 MPa; this is used for chemical plant and any panels subject to atmospheric or marine corrosion.

(*b*) *Heat-treatable:* the treatment in this case depends upon a factor known as *precipitation hardening*; the alloy at room temperature will consist of an α-phase solid solution with a eutectic of α and a metallic compound depending upon which elements are used in the alloying. If the alloy is now heated to approximately 500°C, it becomes a complete solid solution. By quenching from this temperature, the structure is retained, but is obviously unstable because the alloy tries to return to the original α + eutectic. With time at room temperature, the solid solution does in fact decompose into its natural stable structure, and by doing so, changes the mechanical properties from the ductility of the solid solution to the strength and hardness of the α + eutectic structure; this particular change is called *age-hardening*. The precipitation process can be accelerated by heating the alloy to the range 150–200°C and holding it for a few hours before allowing it to cool to room temperature. Both the natural ageing and the artificial process come under the general term of precipitation hardening, whilst the heating to 500°C and quenching to obtain a complete solid solution is known as *solution treatment.*

The advantages of alloys receptive to these treatments are very obvious because, whilst in its ductile state the alloy can be cold-worked, then, by suitable treatment, a tough hard material emerges.

Wrought aluminium alloys which fall into this category are:

(i) Up to 4% copper: these were the original precipitation hardening alloys and are known as the *duralumins.* A metallic compound consisting of one atom of copper to two atoms of aluminium, $CuAl_2$, is precipitated to form the eutectic with the α-phase solid solution. In this condition, strengths in the region of 460 MPa are available. The straight 4% copper duralumins have their limitations and various modifications of this alloy have been developed, particularly to increase strength and stabilize the hardened condition. The corrosion resistance of the duralumins is inferior to the work-hardening wrought alloys and therefore they are often clad with a layer of pure aluminium. Of course, the aircraft industry has been responsible for a lot of the research on these alloys where low weight with strength is at a premium. Rolls Royce Ltd are one firm which has developed a range of aluminium alloys which are known as the R.R.Alloys.

(ii) 1% silicon, 1% magnesium: the metallic compound precipitated in this case consists of two atoms of magnesium to one atom of silicon, Mg_2Si. The treatments involved are the same as for duralumin but the strength of this group of alloys is lower. To compensate for this, the corrosion resistance is higher and they are easier to produce and form; finding application as window sections and structural parts in transport vehicles.

(iii) 6% zinc, 2% magnesium, 1·5% copper: the metallic compound in this case is again $CuAl_2$. Although this range produces the highest strength, up to 620 MPa, it also is the most difficult to produce, and therefore is confined to aircraft and other highly stressed components requiring very high strength : weight ratio.

Cast Alloys

The choice of alloying compositions for the aluminium castings are dependent upon factors other than the resulting strength requirements, e.g. fluidity of the molten metal, soundness of the finished product, and the shrinkage on cooling which should not produce cracking. Elements which are used in aluminium casting alloys are: silicon, to assist the castability; copper and magnesium, or a combination of both, to produce high temperature properties, resistance to corrosion and to provide heat-treatment features.

(*a*) *As-cast alloys:* this group is not subject to any form of heat-treatment, the main alloys being:
- (i) 9–13% silicon: these alloys are approximately of eutectic composition and have the properties of good resistance to marine corrosion, moderate strength (150–230 MPa) and high fluidity making them suitable for intricate castings. A disadvantage of these alloys is their relative brittleness, due to the structure being one of coarse silicon crystals in a coarse eutectic of aluminium and silicon. To reduce this brittleness by producing a fine structure, a process known as *modification* is used; 0·1% sodium is added to the molten alloy restricting the growth of the first crystals formed, and producing a fine *dendritic structure* in a fine eutectic. These alloys find application in sand and die-castings for the automobile and aircraft industries, due to their low weight and high corrosion resistance.
- (ii) 2–4% copper, 5% silicon: the most widely used of the aluminium casting alloys which provides an inexpensive, easily cast material for use where strength is not a prime consideration. It finds application as switch boxes, crank cases etc.
- (iii) 4–6% magnesium, 0·6% manganese: provides the best alloy for corrosion resistance purposes. It has moderate strength (150–230 MPa) and is capable of producing a very high degree of polish. Due to the difficulty experienced in the casting process, it is unsuitable for intricate castings.

(*b*) *Heat-treatable cast alloys:* the heat treatments available in this range are the same as for the wrought alloys, i.e. solution treatment and precipitation treatment and where both these are involved it is termed a full heat treatment. Typical alloys available are:

(i) 10% magnesium: this provides one of the toughest of the aluminium alloys and combines this toughness with a very high corrosion resistance. These properties are desirable in marine and aircraft applications. The components are sand or gravity die-cast and then solution treated by heating to 425°C for approximately 8 hours followed by oil quenching at 100°C. Again, the presence of the magnesium renders casting more difficult.

(ii) 4% copper, 1·5% magnesium, 2% nickel: this group is sometimes referred to as the *Y-alloys* and was developed to meet the need for a material of low weight to resist stresses at moderately elevated temperatures. It can be forged but finds its main use as a casting alloy for such applications as pistons and cylinder heads. It is subjected to a full heat treatment consisting of solution treatment at 515°C for 6 hours and quenching in boiling water followed by a reheat to 100°C for 2 hours.

A guide to B.S.S. 1470–77, 1490

In order to identify the various aluminium alloys a coding system is set out in the above British Standard Specifications. It indicates the form of the material, its composition and any heat treatment to which it can be subjected.

A brief guide to this coding is as follows:

Wrought alloys

An initial letter is used in the prefix to denote which group is being specified:

   N = Non-heat-treatable   H = Heat-treatable

A second prefix letter indicates the form of supply: S, T or E = sheet, tube or extruded section. A number then denotes the alloy composition, followed by a suffix letter indicating the temper in the N alloys and the treatment in the case of the H alloys.

M = as manufactured	H1 to H8 = hardness
O = annealed	H1 being the least hard;
	H8 being fully hard
TB = solution treated	TF = fully heat-treated

Thus an alloy specified as HE 30-TB denotes heat-treatable alloy number 30 in an extruded section which has been solution treated only.

Cast alloys

The code used in this case is the prefix LM followed by the composition number, and a suffix indicating its condition.

No suffix = ingot form   TE = precipitation treated
M = as-cast   TF = full treatment
TB = solution treated

An alloy specified as LM 10-TF is alloy 10 in a cast form which has been solution and precipitation treated.

The full details of the system together with mechanical properties available can be obtained from the standard.

3 Magnesium and its alloys

Magnesium itself is the lightest commercial metal, being approximately two thirds the weight of aluminium. It is chemically very reactive to air and water which means the casting process requires great care and special precautions to protect the melt from the atmosphere. However, magnesium is not sufficiently strong to be used alone for general engineering purposes; it requires to be alloyed with elements such as aluminium, zinc, manganese, and zirconium.

The magnesium alloys have various properties which are attractive to the engineer:

(*a*) Low weight—a property much sought after in aeronautical engineering.

(*b*) Good strength: weight ratio—tensile strengths of up to 300 MPa are available but these alloys possess a low modulus of rigidity which necessitates stiffening for parts subject to any deflection.

(*c*) Machinability—they can be machined at speeds greater than those for aluminium but certain precautions against fire have to be observed. These, however, are well known and by using sharp tools and regularly clearing away swarf, the risk is kept to a minimum.

There are three main alloys available:

(i) 1·5% manganese: possesses a high corrosion resistance, is available in sheet form and is readily welded by inert gas processes.

(ii) Up to 10% aluminium, 0·5–1·0% zinc, 0·3% manganese: available in both cast and wrought forms. The aluminium provides the strength and a metallic compound for precipitation hardening, with the zinc and manganese increasing ductility and corrosion resistance. The heat treatment required is solution treatment followed by precipitation hardening. The increased properties required for the wrought alloys are usually induced by a cold-working process.

(iii) 1·5–5% zinc, 0·6% zirconium: also available in both wrought and cast form with a full heat treatment for the latter. The presence of the zirconium promotes an increase in the number of crystals during the solidification, giving added strength by virtue of a smaller grain size.

Magnesium alloys find many applications both in industrial and in domestic appliances. Examples are large castings such as gear box cases, sumps and airframe components; die-castings for office equipment, domestic cleaners and portable electric tools. In the wrought form they are used as extruded sections for the aircraft industry, ladders and hand trolleys. Generally, a protective coating is provided for magnesium components to assist the corrosion resistance of the surface.

4 Zinc and its alloys

Although zinc is a very weak material it is invaluable to the engineer because of its corrosion resistance properties, brought about by a dense layer of corrosion product which is formed on the surface. It is used chiefly as a cladding or coating for other metals and is applied in many different ways, among them being galvanizing, sherardizing, plating, dipping and spraying, each one having its own particular virtues. Underground pipes and car bodies can be protected by having a block of zinc attached to them. This is known as *sacrificial protection* because the zinc corrodes electrolytically in preference to the iron or steel part to which it is attached.

Apart from being used as a protector, zinc is also alloyed with small amounts of aluminium and copper to provide a very efficient die-casting alloy. B.S.S. 1004 indicates two general die-casting zinc alloys known as A and B:

Alloy A has a composition of 4% aluminium and 0·5% magnesium, giving a tensile strength of 280 MPa. It is used as a general die-casting alloy where dimensional stability is important and it can withstand a certain amount of heat in service.

Alloy B has a composition of 4% aluminium, 1% copper and

0·5% magnesium, with a tensile strength of 340 MPa. It is a harder, stronger alloy than A and is easier to cast but loses impact properties under heat.

Both alloys have one thing in common which is that over a period of 4 to 5 weeks a shrinkage occurs followed by a slight expansion. A stabilizing treatment can be given by heating to 100°C for a few hours and allowing to cool in air.

Zinc die-castings are used extensively in small parts for domestic appliances and lightly stressed car parts, e.g. door handles.

5 Nickel and its alloys

We have already examined the use of nickel as an indispensable alloying element in any alloy steel but it is also extensively used both in its pure condition, and also in conjunction with other elements to produce a range of nickel-base alloys. In its commercially pure form it is used in telecommunication valves and, because of its high corrosion resistance, the chemical industry makes use of it for all types of fittings; food processing is carried out in non-toxic nickel containers. It can be used as the anode in the electrolytic process of nickel plating.

There are many different types of nickel base alloys with varying trade names, the following are perhaps the main types in general use:

(*a*) *Monel:* an alloy of 66% nickel, 33% copper and 1% manganese; it possesses high corrosion resistance with good mechanical properties which are maintained at elevated temperatures. Nickel and copper form a solid solution so that a tough, ductile, corrosion-resistant alloy is formed. The addition of 2–4% aluminium produces a wrought alloy known as K-Monel, which can be precipitation hardened by quenching from 1000°C followed by heating at 600°C for several hours. The tensile strength after this treatment can be as high as 1500 MPa. A casting variety of Monel is obtained by the addition of approximately 4% silicon to the basic Monel. The main industrial applications are found in marine equipment, boiler plant, laundry equipment etc.

(*b*) *Inconel:* an alloy of 76% nickel and 15% chromium, with the balance of iron which is resistant to oxidation and maintains good strength at elevated temperatures; it is used for aircraft exhaust ducts and furnace fittings. It also resists attack by organic substances and is extensively used in food processing and chemical plant. Cronite is another alloy of this general type.

(*c*) *Nimonic alloys:* a range of alloys developed by Henry Wiggin and Co. consisting of varying amounts of nickel and chromium. They have been developed for particular applications in gas-turbine engines where high strength at extreme temperatures and resistance to oxidation are required.

It must be realized that only a few of the many available nickel alloys have been considered and more detailed information on these specialist alloys can be obtained from the various manufacturers.

6 Bearing alloys

Before closing the section on non-ferrous alloys a little time must be spent considering a series of bearing alloys often referred to as the *white metals.*

As we have previously stated in the section covering bronzes, a good bearing consists of hard particles to carry the load, embedded in a soft matrix which wears to provide lubrication channels. The white metals have been developed to meet just this requirement, with two different basic metals being used, i.e. lead-base and tin-base.

(*a*) *Lead-base alloys:* a typical white metal in this range is 80% lead, 15% antimony, 5% tin, and the structure consists of hard crystals of a tin-antimony compound in relatively soft lead-antimony eutectic. They are considered to be inferior to the tin-base alloys but are cheaper to produce and find use in low-duty or light loads such as large fan bearings.

(*b*) *Tin-base alloys:* these are often referred to as *Babbitt metals* and contain 7–10% antimony. The structure is one of hard crystals of tin-antimony compound in a solid solution of antimony in tin. A difficulty in producing these alloys is a tendency for the tin-antimony crystals to gather together instead of dispersing through-out the structure. To prevent this segregation, an addition of 3–4% copper is added which forms a copper-tin compound and solidifies in a needle-like structure before the tin-antimony crystals start to form, so that as solidification proceeds these latter, crystals are entrapped in the network of the copper-tin crystals.

These tin-base alloys are used for high duty applications in automobile and aero engines.

EXERCISES

1. (*a*) Discuss the applications of commercially pure aluminium in general engineering, and state why this material is seldom used in its pure form.

   (*b*) State the effects of (i) copper, (ii) silicon when used as alloying elements in aluminium, and give an example of the application of copper-aluminium alloy and a silicon-aluminium alloy, stating the approximate chemical composition in each case.

   (C.G.L.I.)

2. 'White metal' and certain copper alloys are materials used extensively as bearing metals. State a typical percentage composition for a bearing metal of each material and indicate the service conditions for which each is most suitable.

   (C.G.L.I.)

3. Explain the differences which distinguish a cold-working brass from a hot-working brass.

4. Write brief notes on the following:
   (*a*) stress corrosion;
   (*b*) modification as applied to aluminium-silicon alloys;
   (*c*) hot shortness.

5. Explain fully what the following specification means: HS8-WP.

CHAPTER 4

Sintered Metals

The term sintered metals is applied to the production of certain
small engineering components from fine metallic powders without
the use of a melting process.

In many instances it is the only possible method of production.
The process involves four principal stages:
1. Preparation of the powders
2. Mixing
3. Compacting
4. Sintering
Each of the above stages has to be very carefully controlled to
obtain the desired properties in the final product.

1. Preparation

This stage is concerned with the selection and extraction of the
powders to be used. Certain features of the powders themselves have
to be taken into consideration, as follows:

(a) *Particle size;* this must first be determined in order that
sieving can be standardized. The sizes of the various metallic
powders used must be known because they can affect the density
of the final product.

(b) *Particle shape;* the powders can be of a variety of shapes,
e.g. spheroidal, angular, flat etc. The shape is again an important
factor affecting density and pressing properties. The method of
producing the powder has a relationship to the final particle shape,
e.g. chemical precipitation, condensation, crushing or grinding will
each give a different particle shape; therefore the shape required

must be determined and the method of powder production made to suit.

(*c*) *Density;* the density of each particle must obviously have a bearing upon the final product density. A particle can have porosity in two general forms:

    (i) continuous pores running through it (Fig. 7A);
    (ii) isolated pores which are not connected (Fig. 7B).

Porosity in particles is important from the viewpoint of surface area accessible to the oxygen of the atmosphere. Metal in a finely-divided form is more chemically active than when in a large form, since it has more surface exposed to the atmosphere. It can be seen from Fig. 7 that particles with interconnected pores present more surface area than particles with isolated pores. Further, greater porosity allows greater absorption of atmospheric gases, resulting in uneven distribution of pressure during compacting which produces an unstable stressed condition.

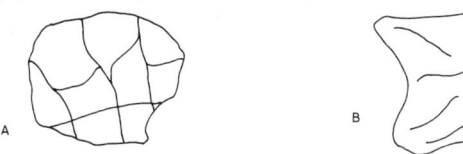

Fig. 7 *Types of pores in powder particles A. With continuous pores.*
B. With isolated pores.

(*d*) *Chemical purity;* this needs to be of the order of 99 % minimum and most industrial metals are obtainable in this pure form.

2. Mixing the powders

In the mixing stage, it is important that an even distribution of the various powders is obtained. One method of mixing is known as ball-milling and consists of rotating the powders together with loose steel balls in a drum. By this method one metal is smeared on to particles of another; this assists the sintering stage.

3. Compacting

After mixture, the powders are carefully weighed, placed in a steel die and subjected to pressures of the order 600–800 MPa. The die may be of the single-punch type, where the bottom is fixed, but more usually it will be double-acting, where top and bottom dies move towards each other.

4. Sintering

This is a heating process in which the compacted powders are baked or fused together to give strength and hardness to the final product. It is not a melting process even though a small amount of melting may occur. Sometimes the process is performed in two stages: pre-sintering and final sintering. After pre-sintering, the compacts can be cut or machined to form and size, as is usual in the manufacture of hard carbides where, after final sintering, no further machining, but only a polishing process is possible.

The temperatures and times involved in the sintering stage are closely controlled because of the shrinkage which can occur. When a single metal is used the temperature is usually of the order of three-quarters of the melting temperature; for a mixture it is the melting temperature of the lowest-melting-point metal.

Furnaces are usually electrically heated, with the atmosphere controlled by the use of an inert gas.

Types of Sintered Products

Porous bearings. The main advantage of porous bearings is that they can be impregnated with oil to give them self-lubricating properties. Porous bronze bearings are quickly manufactured, due to the short time required for sintering; they have a very low coefficient of friction, high wear resistance and a strength of 70–140 MPa. Popular bearing materials are the iron-based alloys, which are strong and have the same coefficient of friction as the steel shaft which they support.

Non-porous bearings. When a heavy-duty bearing is required, a non-porous copper-lead or copper-tin-lead mixture is sintered on to a steel backing strip. The whole is rolled to a theoretical density and then annealed. A better distribution of the lead particles is thus obtained. When the bearing has to run without oil or grease as a lubricant, polytetrafluorethylene (P.T.F.E.), a plastics material, is held in a porous metal matrix, which supports the plastic and conducts the heat from the bearing. The low strength and poor conductivity of P.T.F.E. prevents it from being used alone as a bearing but it possesses the advantages of stability up to 300°C under almost all corrosive and oxidizing conditions with a low coefficient of friction.

Filters. Sintered products can be produced in a large variety of shapes and the pore size can be closely controlled. These facts make sintered metals ideal for filters and diaphragms. Nickel, Monel and stainless steel filters are used for filtering fuel oils, liquid chemicals and emulsions, and are ideal for separating mixed liquids of varying surface tensions. A successful application is the separation of water from fuel in a jet-engine fuel-supply system.

Friction Materials. Sintered metals are now being used as friction components on heavy-duty applications, e.g. aircraft brakes and heavy machinery clutch and brake linings. They consist of a continuous metal matrix into which are bonded non-metallic friction generators. Resistance to wear is excellent and a wide range of friction characteristics can be obtained.

Structural Parts. The mass production of components is possible with powder metallurgy, but invariably the stresses which the parts have to withstand and the inherent porosity of the sintered parts are undesirable for such productions. By various techniques, e.g. double-pressing and sintering and by adding graphite, nickel, manganese and molybdenum, strengths have been improved.

Electrical and Magnetic Components. The composite structures available by means of powder metallurgy are used for electrical contacts and current-collector brushes. The low contact resistance of silver and copper can be combined with the strength, heat resistance and arc-erosion resistance of tungsten, molybdenum and nickel or the lubricating properties of graphite. Sintered metals can be made into magnets, pole pieces, armatures and cores for self-induction coils for radio and television parts.

Hard Metals: the cemented carbides. This range of metals is the hardest of any known metal or alloy. They can be composed of cobalt, chromium, molybdenum, titanium, tantalum or tungsten carbides. The material can be heated to red or orange heat without losing its hardness and so is suitable for cutting tools used at high speeds and temperatures.

The sintered carbides cannot be forged or cast and can only be shaped. after the forming process by grinding. They are very expensive, so they are supplied in the form of tipped tools. These tools are suitable for cutting any metals and such materials as glass, bakelite, ceramics, fibre, ebonite, plastics etc, which would dull the edge of ordinary tools. Extremely hard and hard wearing, the carbides are used for such applications as lathe centres, milling cutters, blades, drills, tool tips, wire drawing dies, jaws of inspection gauges, and the wearing surfaces of plug and ring gauges.

There is a large number of proprietory brands of tungsten carbide tools on the market. Each manufacturer gives recommendations of the grade of tool to be used for machining particular materials and it is suggested that these should be perused.

Advantages of the Sintering Process

 (*a*) components having specialized properties can be produced, i.e. self-oiling bearings;

 (*b*) it can be used to unite metals which, for various reasons, do not alloy together or have widely differing melting points. It is also suitable for combining metals and non-metals;

 (*c*) it enables such metals as tungsten which have an unusually high melting temperature but possess desirable properties, to be used without recourse to highly complex casting precautions;

 (*d*) it lends itself to the mass-production of finished components;

 (*e*) very little space is required to utilize the process.

Disadvantages of the Sintering Process

 (*a*) expensive metal powders;

 (*b*) the size of a component produced by this method is very limited;

 (*c*) the dies are costly and therefore a large production run is essential to justify this initial cost;

 (*d*) the design of the product has to be suited to the pressing operation.

EXERCISES

1. Compare powder metallurgy with the other forming processes of casting and forging.

2. Describe with aid of sketches how a brush would be produced by the powder metallurgy process.

3. Write brief notes on four factors to be considered when designing components for the process of powder metallurgy.

4. Describe fully any one process whereby a metallic powder is obtained.

5. Describe how the lubricant is installed in an oil-impregnated bearing.

CHAPTER 5

Plastics

Definition

The term 'plastic' is applied to any material which can be shaped when a force is applied to it, i.e. the force causes the material to deform into a predetermined shape.

When the term 'plastics' applies to a range of materials it refers to a chemically manufactured group of materials which generally have an organic base and are in a plastic condition during the shaping processes.

Therefore the materials used are manufactured rather than obtained naturally like ores, wood and rock. The term organic is applied to chemicals in which carbon is a basic material.

Introduction to terms used

1. Monomer: this is any substance which is composed of molecules and constitutes the basis for the whole range of synthetic resins and plastics. These monomers are capable of joining together to form a chain formation.

2. Polymer: when a series of monomers have been chemically induced to join together in a chain formation or one large molecule, the resulting larger molecule is called a polymer. It can consist either of one monomer, or two or more different monomers combined together. The process is known as polymerization, and the aim is to produce as large a chain as possible, giving rise to the term high polymer. The larger the chain or molecule the better properties are induced in the plastic.

3. Copolymer: refers to a structure in which different monomers have been combined and subjected to a polymerization process.

4. Natural polymers: although most polymers are artificially produced there are, in fact, several natural polymers:
   (a) *cellulose*, a naturally occuring polymer of the grape sugar, glucose;
   (b) *casein*, obtained from skimmed milk;
   (c) *bitumen*, which undergoes no chemical alteration when used in plastics.

5. Catalyst: the polymerization processes are chemical reactions and as such are capable of being accelerated by the addition of other chemicals. These other chemicals are known as catalysts and simply increase the reaction, playing no other part in the final product. When a violent chemical reaction occurs then negative catalysts are used to slow it down.

6. Plasticizer: this is a chemical which is added to the plastic to remove brittleness and lower the softening temperature. It should be compatible with the resin material or it will sweat out from the finished product either as a bloom or as an oily smear.

With *thermo-set plastics*, the plasticizer acts as a flux to enable the material to flow more easily in the mould. Furfural is sometimes used for this process and after performing this function is absorbed by the plastic in a chemical combination.

P.V.C. when hot-pressed alone is hard and brittle with a softening temperature too high for easy hot-working. By using a plasticizer or blend of plasticizers, the brittleness is removed and the softening temperature lowered; a tough, pliable material of easy flow is produced.

7. Fillers: are added to plastics to give an increase in properties of strength and electrical resistance, or they are added simply to provide bulk. Inexpensive, low strength plastics can have 75% or more of a filler. Some of the common filler materials are:
   (a) *calcium carbonate* which provides bulk alone;
   (b) *wood flour* which provides bulk with reasonable strength but tends to absorb moisture;
   (c) *mica* which increases electrical resistance with low moisture absorption;
   (d) *cloth and fibres* which give strength but add to the cost;
   (e) *asbestos fibres* which increase heat resistance.

Types of plastics

There are mainly two classifications applied to plastics as follows:

1. Thermo-Set Plastics

This range requires both heat and pressure to mould them to shape. They are distinguished by the fact that on heating they become soft and pliable, but if the heating is continued a chemical change occurs which causes them to harden. They are permanent at this stage and if re-heated they will not soften; if the heat is sufficient however, it will eventually cause a breakdown of the material. The hardening of these plastics is often referred to as *curing*.

Important thermo-set plastics are:

(*a*) *Phenolics:* manufactured from the chemicals phenol and formaldehyde. In the production of this range a resin compound is made from the two chemicals and mixed with a suitable filler material producing either a powdered or a flaky compound depending upon the type of filler used. By application of heat and pressure, a plastic mass is produced to the shape of the mould prior to the curing process.

Laminated plastics of this type are produced by dissolving the phenol-formaldehyde resin in a suitable solvent and then passing the solution over cotton or asbestos cloth. The solvent dries off, leaving the resin behind attached to the fabric sheet, of which several layers are pressed together producing a strong laminated sheet. They are important for components requiring electrical resistance with high strength.

(*b*) *Amino-plastics:* manufactured from the chemicals urea and formaldehyde. They are paler and more light stable than the phenolics but are not as resistant to moisture.

(*c*) *Polyester:* this is a thermo-set resin, commonly bonded with fibre-glass, which has been woven into a cloth. It has found use as a material for automobile bodies, small boats, protective helmets and several highly stressed aircraft parts.

(*d*) *Epoxides:* a group of resins known as the *epoxy* resins form the most useful material in this range. Araldite is a well known form of this plastic and is used extensively as an industrial adhesive. It can be hardened by addition of a chemical, or it can be obtained in a hot-setting form which is cured in the usual manner of heat application. Epoxides have the properties of good chemical resistance and low shrinkage. One application, besides as an adhesive, is in the form of moulds or templates used in copy-milling and stretch forming processes.

2. *Thermo-Plastic Materials*

This group of plastics softens on the application of heat but requires cooling to harden permanently to shape. Any further application of heat will simply re-soften the material because no chemical action has taken place.

There are many types of this plastic available and amongst the more important are:

(*a*) *Nitro-cellulose:* this was the first thermo-plastic to be developed, under the name celluloid. It is tough and flexible but has the disadvantage of being inflammable. It is used in the form of fibres, sheets, rods and tubes.

(*b*) *Cellulose-acetate:* in this type the cellulose is combined with acetic anhydride to give a similar product to the nitro-cellulose range but it is non-inflammable. It loses a little in toughness but is widely used for packaging, tool coating, steering wheels etc.

(*c*) *Polyvinyl-chloride* (*P.V.C.*): this is manufactured from calcium carbide and acetylene. It provides a tough, non-inflammable material widely used as a floor covering and, because of its resistance to oil and water, it is used as guttering and wire covering.

(*d*) *Polystyrene:* made from pure benzene. This is lighter than most other plastics but has a tendency to be brittle; it has good electrical resistance and is used as an insulator for high frequency parts, battery, radio and television cases. It can be moulded, cast or formed to the desired shape.

(*e*) *Polyethylene:* commonly known as polythene, is manufactured by the polymerization of ethylene. It is resistant to alkalies, and acids and is tough and flexible, which makes it a good insulator. It is used in wire covering, piping for plumbing and chemical plant. Flexible polythene bottles are common features of the domestic scene. A further development of this plastic is polypropylene which has a higher softening point and is used for items such as distributor caps in the automobile industry.

(*f*) *Polytetrafluorethylene* (*P.T.F.E.*): consists of atoms of fluorine and carbon. It has high chemical and electrical resistance, with outstanding features of a low co-efficient of friction so that other substances will not adhere to it. It is costly and finds use mainly in chemical containers and seals.

(*g*) *Polymethyl methacrylate:* commonly known as perspex, is made from acetone and prussic acid. It has high transparency with good strength and moisture resistance. Applications include knobs, dial shields and electrical parts.

(*h*) *Polyamides:* commonly known as nylon, are a range of tough wear resistant materials which have many engineering applications such as gearwheels and bearings.

It can be seen from the many variations of plastics that their utilization as engineering materials is still in the development stage. The production of the raw material in a form usable to the engineer is of course, the domain of the chemist. However, its application and development is very much in the hands of the mechanical designer.

CHAPTER 6

Loci and Cams

Loci

The locus (pl. loci) of a point is the path traced out by that point as it moves under the conditions of restraint imposed upon it.

Often when the conditions of restraint are mathematical laws, the loci traced out are common geometrical forms.

The circle is traced out by a point which moves so that its distance from a fixed point is constant (Fig. 8A).

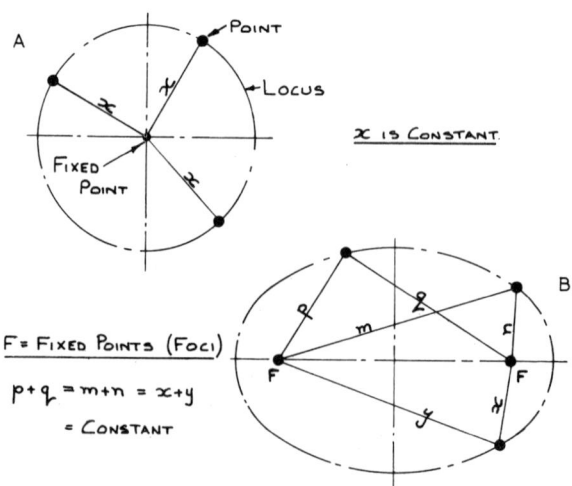

Fig. 8 Geometric loci

The ellipse is traced out by a point which moves so that the sum of its distances from two fixed points, called focal points, is constant (Fig. 8B).

When mechanisms are being designed, it is sometimes essential that the locus of a point on the mechanism be determined so that clearances can be checked, forces in linkages be determined, and correctly shaped machine guards be designed.

The locus should be plotted by drawing the mechanism in several positions and determining the position of the tracing point for each position of the mechanism. Many mechanisms have a revolving crank as the driving member and it is convenient to take 12 positions of this crank as the basis for fixing the positions of the mechanism.

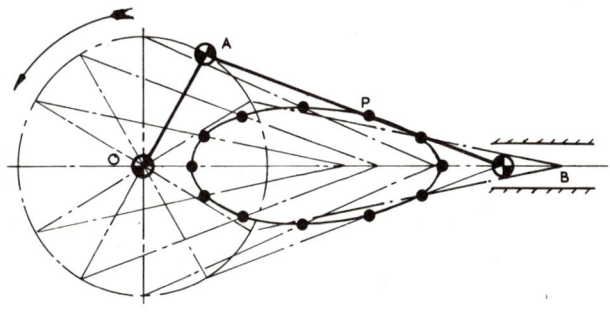

● Pivot Point

● Locus Point

Fig. 9 Locus of a point on a connecting rod

Example 1

Fig. 9 shows a crank OA rotating about a centre, O, at constant speed. AB is a connecting rod, fastened at B to a sliding member constrained to reciprocate along the centre line BO. It is required to draw the locus of the point P. OA is drawn in 12 positions together with the 12 positions of AB. The point P is found for each position of AB and a smooth curve is drawn through these points. The curve is the locus of the point P.

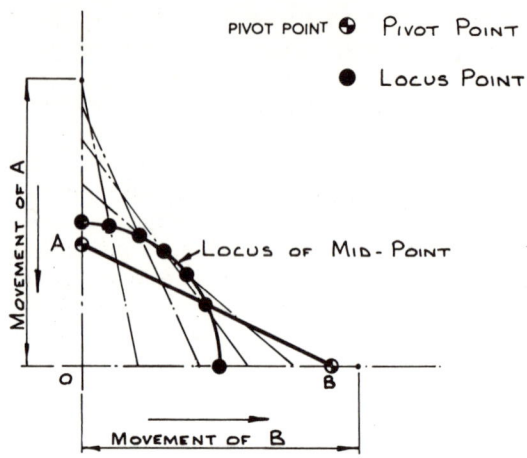

PIVOT POINT ⊕   Pɪᴠᴏᴛ Pᴏɪɴᴛ

● Lᴏᴄᴜѕ Pᴏɪɴᴛ

Fig. 10 Locus of mid-point on a sliding bar

Example 2

Fig. 10 shows a link AB connecting two sliding members A and B. In operation, A slides towards O as B slides away from O, and vice-versa. The locus of the mid-point of the connecting rod is required whilst A moves from its extreme outer position to O.

In this case the positions for fixing the mechanism can be selected at random. When the rod AB has been fixed in all positions, its mid-points can be determined and its locus drawn in.

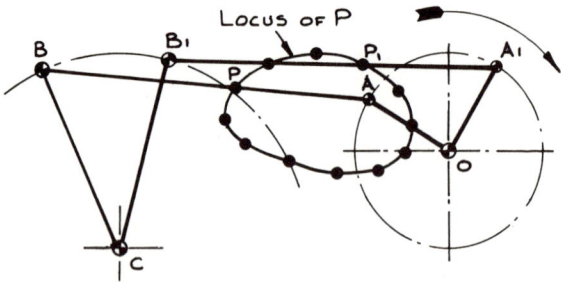

⊕ Pɪᴠᴏᴛ Pᴏɪɴᴛ

● Lᴏᴄᴜѕ Pᴏɪɴᴛ

Fig. 11 Locus of a point on a connecting rod

Example 3

Fig. 11 shows crank OA rotating about O at constant speed. The connecting rod AB is fastened to a swinging link BC, pivoted at C. the locus of point P is found by fixing the mechanism in 12 positions.

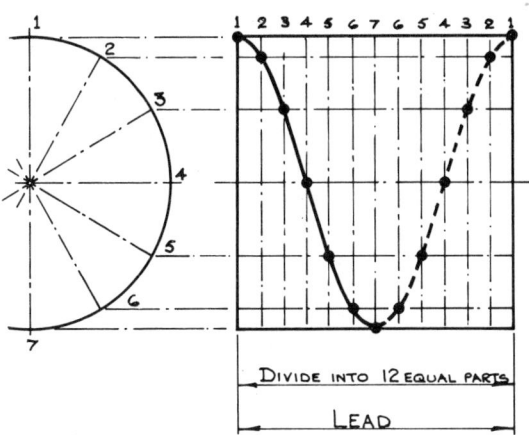

Fig. 12 The helix as a locus

An important locus is the helix. This is the locus of a point which moves around the circumference of a cylinder at the same time moving axially. The ratio between the two movements is constant.

Fig. 12 shows the construction of a helix. The axial movement corresponding to one revolution of the circumference is known as the lead.

The screw thread is a helix and helical springs take their name from the helix. The construction of a square thread is shown in Fig. 13.

Many mechanical operations require motion other than that which is supplied. Most machine tools are supplied with circular or rotating motion from an electric motor and convert this to reciprocal or straight line motion by some means.

Consider a lathe tailstock which is fitted with a screw thread. Rotation of the handwheel is translated into straight line motion of the barrel through a helix. This type of translation is common. It occurs in the lead screws and cross slides of machine tools, in measuring instruments, e.g. the barrels of micrometers, and in providing the force in a fly-press, screw-jack or vice.

49

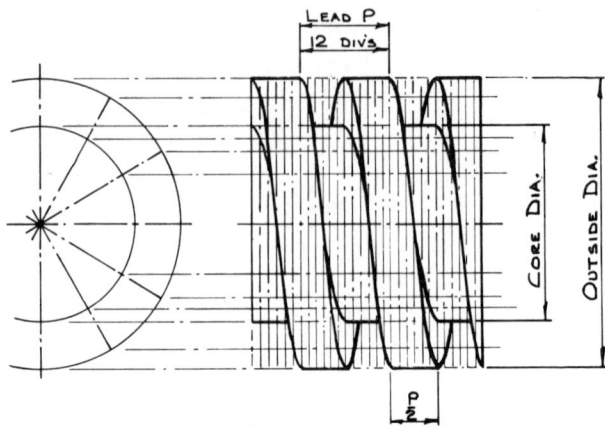

Fig. 13 Single start, right hand square thread

In petrol and steam engines, reciprocal or straight line motion is converted into rotational motion through a crank. The action in Fig. 9 is similar to that experienced in a petrol engine, but in this case the sliding member, or piston, causes the crank to rotate. In steam engines and locomotives a crosshead is employed in the translation (Fig. 14).

Conversely, straight line motion is obtained from rotational motion in machine hack-saws, shaping machines and slotting machines.

Another method of obtaining straight line motion from rotational motion is through a rack and pinion. A planing machine table is actuated by a pinion driving a rack; the saddle of a lathe is similarly

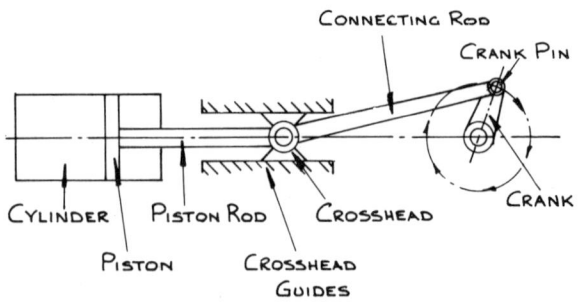

Fig. 14 Locomotive crosshead device

moved along the bed, but in this case the bed and rack are stationary.

Again, this system can be used in measuring instruments. The plunger of a dial gauge pushes a rack over a pinion, which, through gearing, actuates the needle.

A non-reversible system of translation from rotational motion to straight line motion is by means of an eccentric. This system is often employed to drive small lubrication pumps or number counters on machinery having a continuous operation. The eccentric is also used as the motion translator on power presses.

Cams

A cam is a body having a special profile which imparts reciprocating or oscillating movement to another body, called a *follower*.

The profile of the cam depends upon the cam's own movement, the shape of the follower and the desired movement of the follower.

They are divided into two important types:

(a) *Radial, edge, plate or disc cams*

In this type of cam, the follower oscillates or reciprocates in a plane at right angles to the cam axis and the working edge of the cam.

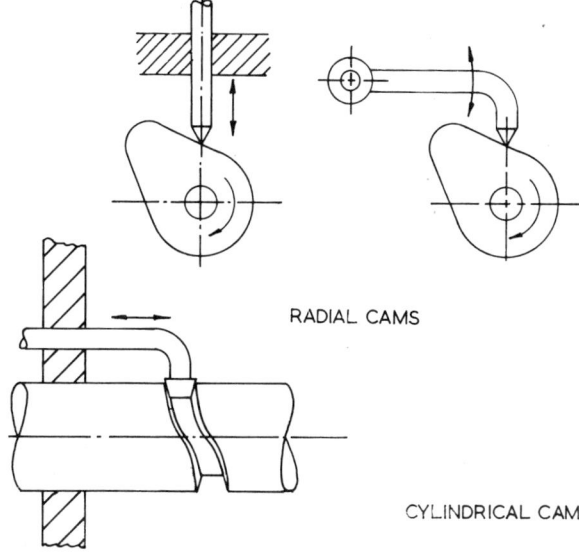

RADIAL CAMS

CYLINDRICAL CAM

Fig. 15 Types of cam

51

(*b*) *Cylindrical cams*
In this type of cam, the follower oscillates or reciprocates in a plane parallel to the cam axis.

Fig. 15 shows the different cams with reciprocating and oscillating followers. Radial cams only will be dealt with in detail as these constitute the greater number used in practice.

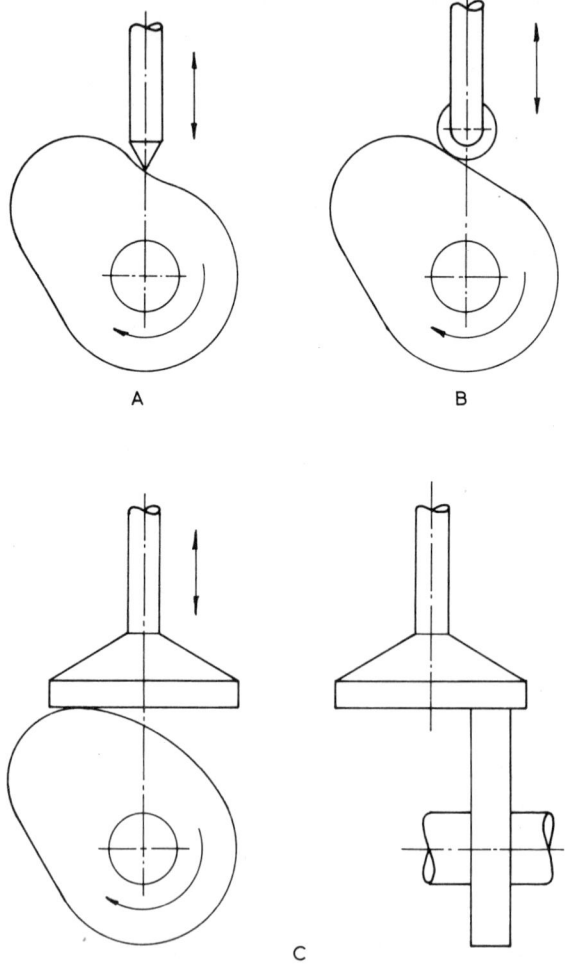

Fig. 16 Types of cam follower

Cam followers

Cam followers may have oscillating or reciprocal motion and this, together with the shape of the cam, will determine, to some extent, the type of follower used.

(*a*) *Knife, wedge or point follower.* With this type of follower, the rate of wear at the contact point is too great for them to be in common use, but they can be used with cams of any shape (Fig. 16A).

(*b*) *Roller follower.* Wear is almost eliminated with this type of follower as there is little sliding between the contact surfaces. The radius of the roller must be smaller than the least radius in the cam profile (Fig. 16B).

(*c*) *Flat or mushroom follower.* Considerable side thrust between follower and guide is experienced with knife and roller followers. Side thrust on the guides, using a flat follower, is due solely to sliding friction between contact surfaces of cam and follower. Wear and side thrust on the flat follower can be reduced if the axis of the follower is offset. As the cam rotates, friction causes the follower to rotate on its own axis (Fig. 16C).

When using a flat follower, the working edge of the cam must be entirely convex.

The contact between the cam and its follower is positive only on the outstroke and springs must be fitted to keep both in contact for the remainder of the cam rotation.

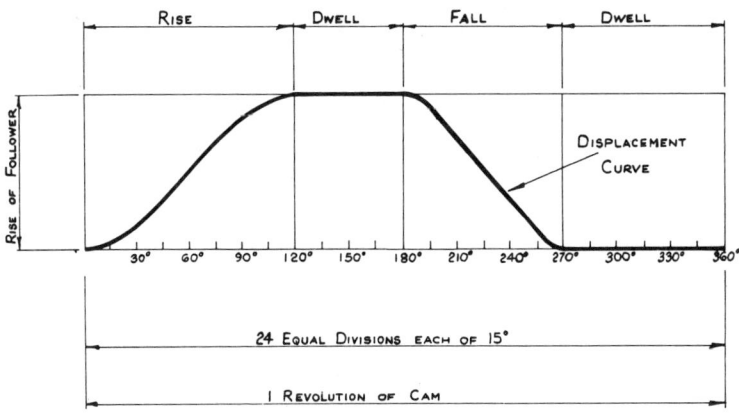

Fig. 17 Typical displacement diagram

Follower Motions

In the construction and study of cams a displacement diagram or cam graph is used. This shows the displacement or lift of the cam follower plotted against the angular displacement of the cam. The graph is plotted for one complete revolution of the cam.

The cam usually rotates at constant speed so that equal amounts of angular displacement occur in equal intervals of time. The follower, however, is required to start and finish its stroke at rest, thereby having variable velocity. Further, it often remains at rest during some part of the cam rotation; this is known as a period of dwell.

The displacement curve shown in Fig. 17 depends on the type of follower motion required, but it represents a typical displacement curve. The types of motion are as follows:

1. Uniform velocity (see Fig. 18A). If the follower is to have uniform velocity, it must have equal amounts of lift for equal amounts of angular displacement, the slope of the curve is constant, i.e. a straight line. In practice, the severe changes in direction at the

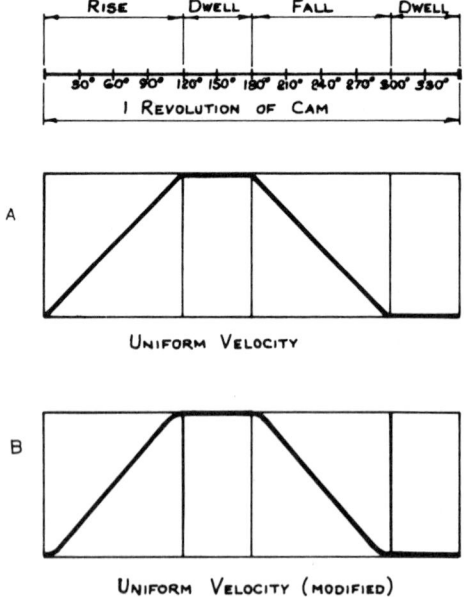

Fig. 18 Uniform velocity displacement diagrams

start and finish of the motion are smoothed out by radii, preventing the follower from jumping on the cam profile (Fig. 18B). When a uniform velocity cam is to be constructed, the modification to the displacement diagram can be ignored as the smoothing will be done when the profile is sketched in.

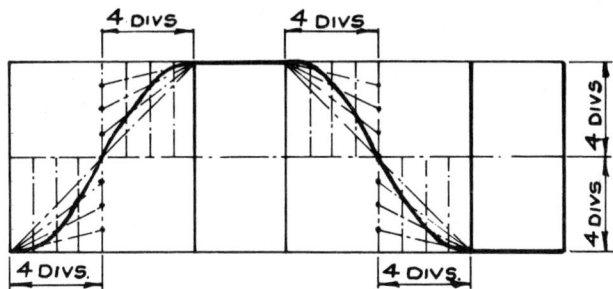

UNIFORM AND EQUAL ACCELERATION AND RETARDATION

Fig. 19 Uniform and equal acceleration/retardation displacement diagram

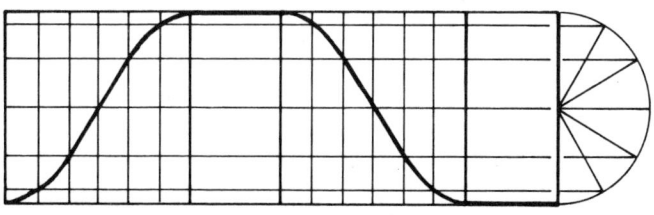

SIMPLE HARMONIC MOTION

Fig. 20 Simple harmonic motion displacement diagram

2. *Uniform and equal acceleration and retardation (see Fig. 19).* The curve for this type of motion is of parabolic form. The follower accelerates for the first part of the motion and retards during the second part. The lift of the follower and the angular displacement are divided into the same number of equal parts and the method of construction for the parabolic curves is shown in the diagram.
3. *Simple harmonic motion (see Fig. 20).* When the follower has simple harmonic motion the curve of the displacement diagram is a sine curve. Construction is similar to that of the helix (see Fig. 12).

55

Figs. 21, 22 and 23 show the construction of cams having different types of followers and various motions. Our study is restricted to followers having their lines of action on the cam centre line.

PLATE CAM DESIGNED TO GIVE THE FOLLOWING MOTION TO A WEDGE FOLLOWER :-

 0° - 150° RISE OF 30mm WITH UNIFORM VELOCITY.
150° - 180° DWELL.
180° - 300° FALL OF 30mm WITH SIMPLE HARMONIC MOTION.
300° - 360° DWELL.

LINE OF ACTION THROUGH CAM CENTRE. ANTICLOCKWISE ROTATION.
LEAST RADIUS OF CAM - 30mm.

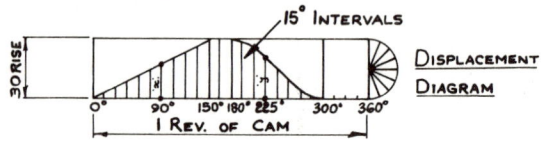

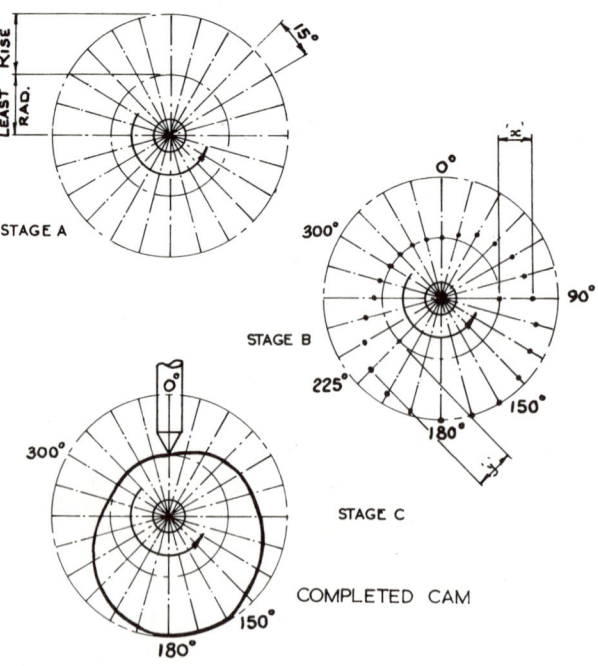

Fig. 21 Plate cam design with wedge follower

PLATE CAM DESIGNED TO GIVE THE FOLLOWING MOTION TO A 25mm DIA
ROLLER FOLLOWER :-

    40mm LIFT IN 150° WITH UNIFORM ACCEL'N./RETARD'N.

    DWELL FOR 60°

    40mm FALL IN 150° WITH SIMPLE HARMONIC MOTION.

LINE OF ACTION OF FOLLOWER THROUGH CAM CENTRE.

    ANTICLOCKWISE ROTATION

    LEAST RADIUS OF CAM - 25mm.

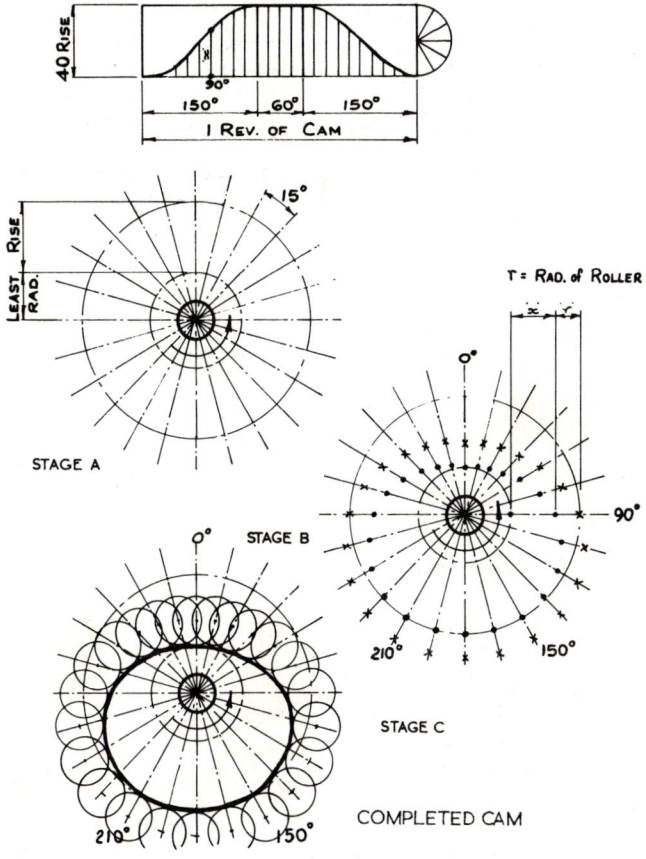

Fig. 22 Plate cam design with roller follower

57

PLATE CAM DESIGNED TO GIVE THE FOLLOWING MOTION TO A
FLAT FOLLOWER :-
      30mm LIFT FOR 120° OF CAM ROTATION WITH UNIFORM.ACCEL/RET'N
      DWELL FOR 60°.
      15 mm FALL FOR 60° WITH UNIFORM VELOCITY.
      DWELL FOR 60°.
      15mm FALL FOR 60° WITH UNIFORM VELOCITY.

LINE OF ACTION THROUGH CAM CENTRE ANTICLOCKWISE ROTATION

    LEAST RAD. OF CAM - 30 mm.

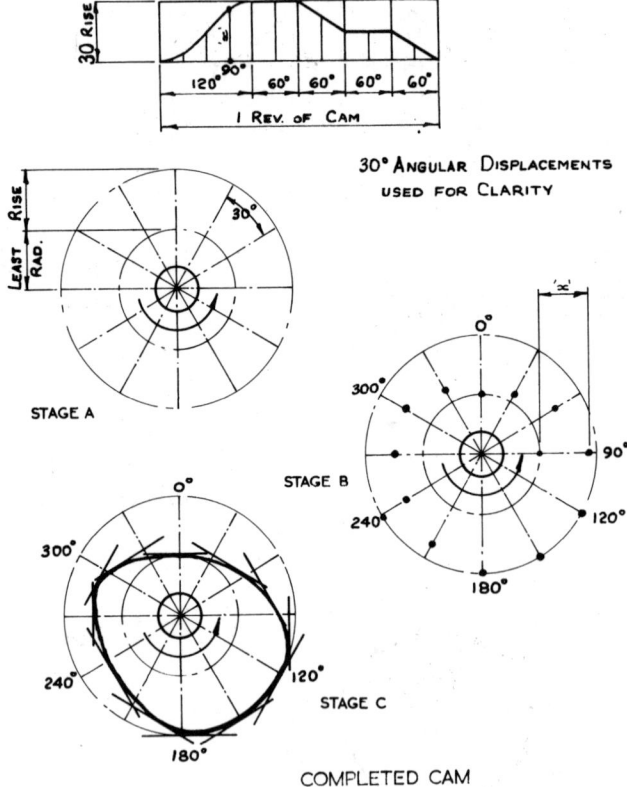

Fig. 23 Plate cam design with flat follower

The Construction of Cam Profiles

In each case the displacement diagram for the cam follower is constructed first.

(*a*) Fix the cam centre and determine the nearest and farthest position of the cam follower. Draw circles through these points and divide the circles into 15° increments.

(*b*) Step off, from the displacement diagram, the amounts of follower lift and transfer to the corresponding radial line on the cam diagram. Remember that if the cam is to have anti-clockwise rotation, the follower displacements should be stepped off clockwise, because the cam is stationary.

(*c*) Complete the cam by drawing in the profile and adding any other necessary information.

The above system would be used with a wedge or knife follower as the points stepped off correspond to the point of contact of the follower. When a roller or flat follower is used, the point stepped off is used for the construction of the shape of the follower. When all the followers have been drawn in the cam profile is drawn tangent to the followers, as shown in Figs. 22 and 23.

Circular Arc Cams

The cams which have so far been considered are difficult and costly to make. The uniform velocity cam is the only one which can be

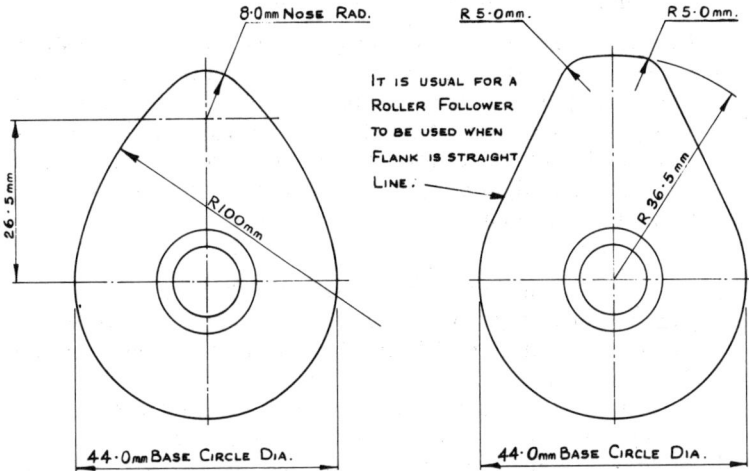

Fig. 24 Circular arc cams

generated on a machine tool (milling machine). The other two have to be produced largely by hand. These are then used as templates for the production of other cams of the same profile.

Circular arc cams are cheaper to produce and are likely to be more accurate, with their profiles composed of circular arcs and straight lines. A typical application of this type of cam is for operating the valves of small internal combustion engines. The displacement curve of such a cam is often very similar to that of a simple harmonic motion or acceleration/retardation curve and differs only by small amounts. The construction of circular arc cams is shown in Fig. 24.

EXERCISES

1. The crank OA of the mechanism shown in Fig. 25 revolves clockwise at a constant speed about O. The end B of the link PAB slides along XY. Determine the locus of the end of the link P if OA is 25 mm; AB is 65 mm; and AP is 50 mm.

2. In Fig. 26, the crank OA of the mechanism shown, revolves anticlockwise at constant speed. The end B of the link ACB slides along XY. To this link is connected another link CD which slides through the pivoted block at E. Draw the locus of the point D for one revolution of OA. OA is 40 mm; AB is 125 mm; BC is 50 mm; CD is 100 mm.

3. The crank OA of the mechanism shown in Fig. 27 revolves clockwise at constant speed about O. In one revolution of the crank, the point A moves from its initial position to O and back again. Two links AB and BC are pivoted at B and connect A and C. In one revolution of the crank, point C moves to D and returns, both at constant speed. For one revolution of the crank, plot the locus of the pivot B. OA is 380 mm; AB = BC is 1070 mm.

4. Fig. 28 shows the outline of a transfer mechanism. AB and CD are arms pivoted at A and C. E rotates about P and is connected to the arms by EF and EK. Suction devices at B and D convey work to the operating point O.
   (a) Find by construction, the radius of rotation EP required to bring B to O at the extreme swing of the arm.
   (b) Make sketches to show how the mechanism must be arranged to fulfil the following conditions:

(i) for **B** and **D** to meet at **O**;

(ii) for the outward swing **OB** to be quick return action.

(C.G.L.I.)

5. Draw two complete leads of a left hand, two start square thread having 125 mm outside diameter and 75 mm lead.

6. Fig. 29 shows the cross section of a sheet metal chute of negligible metal thickness. The chute is in the form of a helix (R.H.) with a mean diameter of 3 m and a lead of 2 m. Draw the plan and elevation of *one* complete lead of this chute.

7. Make a neat sketch of the linkage mechanism employed in a shaping machine. Explain why it can be described as a 'quick-return motion'.

8. Describe, with the aid of sketches, two methods by means of which a roller follower can be kept in positive contact with a plate cam.

9. Construct a cam profile to give the following motion to a roller, 25 mm diameter:
   dwell for 90° of cam rotation;
   rise 40 mm with simple harmonic motion for 75° of cam rotation;
   dwell for 30° of cam rotation;
   fall 40 mm with uniform velocity for 75° of cam rotation;
   dwell for remainder of revolution.
   Least radius of cam 50 mm; anticlockwise rotation. Line of action through cam centre.

10. Construct a cam profile to give the following motion to a flat follower; rise and fall of 25 mm to take place with uniform acceleration/retardation. Rise of follower in 90° of cam rotation, dwell for 10°, fall in 90° and dwell for remainder. Least radius of cam 40 mm. Clockwise rotation. Line of action through cam centre.

11. Draw full size, the profile of a plate cam rotating clockwise which gives a 20 mm diameter roller follower the following motion:
    0–120° lift 25 mm with uniform acceleration and retardation
    120–180° dwell
    180–240° lift 15 mm with uniform velocity
    240–360° fall 40 mm with simple harmonic motion.

12. Fig. 30 shows the profile of a circular arc cam. This cam operates an in-line roller follower 20 mm diameter. If the cam rotates in an anticlockwise direction, determine the displacement curve for the follower. Assume the position of the cam to be in its starting position.

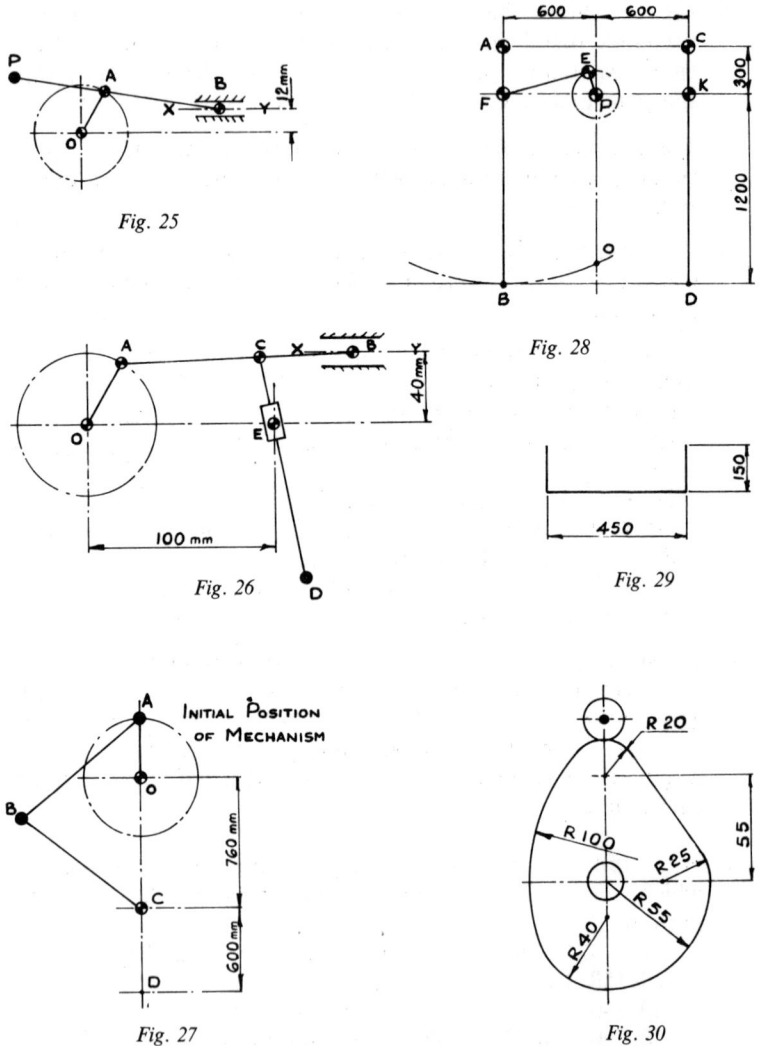

Fig. 25

Fig. 28

Fig. 26

Fig. 29

Fig. 27

Fig. 30

CHAPTER 7

Gears

Toothed gearing has been continuously developed since wooden pegs were fitted into wooden wheels and meshed with wheels with slots cut in them. Plain cylinders, transmitting power by friction, have also been used in the past and the principle still finds many applications.

Modern gears use a tooth form known as *involute*. The side of each tooth is curved to this form by means of modern gear-tooth generating machines. The involute curve is generated by a straight line rolling around a circle without slipping; any points on the straight line trace out involute curves as the line rolls. Fig. 31 shows the construction of an involute curve.

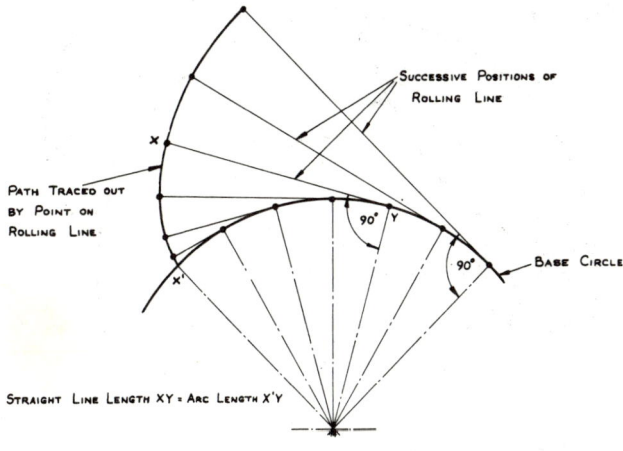

Fig. 31 Construction of involute curve

Prior to the advent of machine-cut gears, gear teeth, when in *cycloidal* form, were cast integral with the gear wheel. The cycloid is a curve traced out by a point on the circumference of a circle which rolls without slipping along a straight line (Fig. 32A).

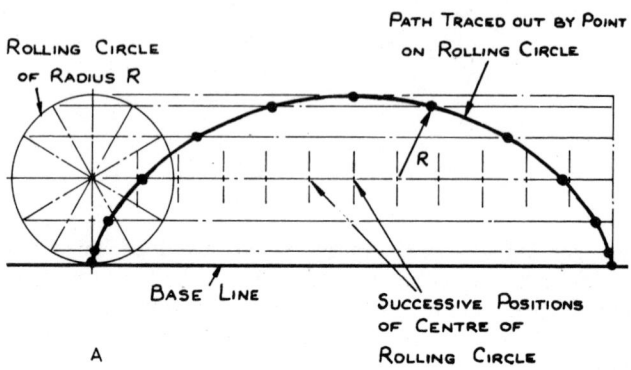

A

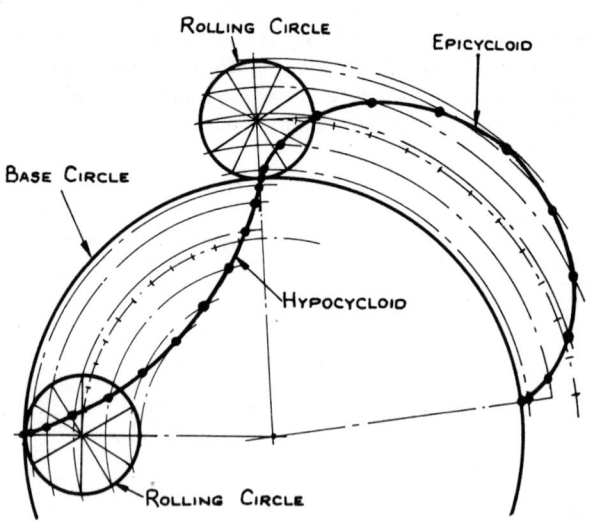

B

Fig. 32 Construction of cycloidal curves A. Cycloid curve. B. Epicycloid and hypocycloid curves

When the circle rolls, again without slip, on the outside or inside of a larger circle, the loci traced out are called *epicycloid* or *hypocycloid* respectively. Epicycloid and hypocycloid curves are shown in Fig. 32B. The cycloidal tooth form is that portion of the epicycloid and hypocycloid where the curve crosses the base circle. A comparison of the two tooth forms is shown in Fig. 33. The cycloidal tooth form is difficult to produce by machine tools and the involute form has become accepted as standard.

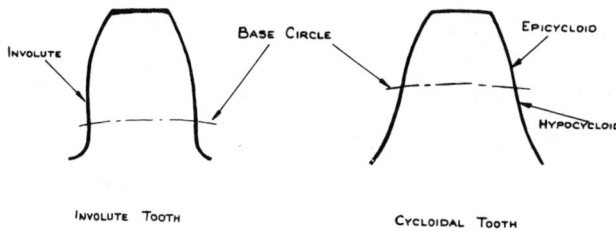

Fig. 33 Types of gear teeth

There are several forms of gearing, the most common being the spur gear (Fig. 34), which will transmit power between two parallel shafts lying in the same plane. Fig. 35 shows bevel gears, the teeth of which are cut on a cone. These will transmit power between two shafts lying in the same plane, but whose axes would intersect if

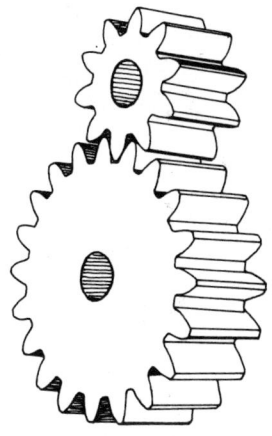

Fig. 34 Spur gears in mesh

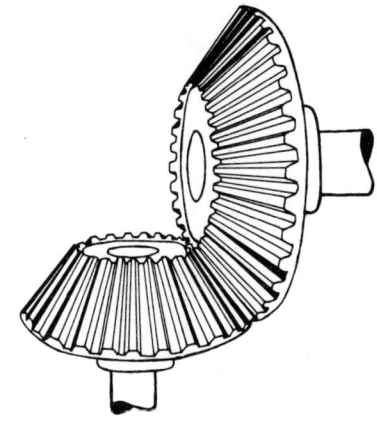

Fig. 35 Bevel gears in mesh

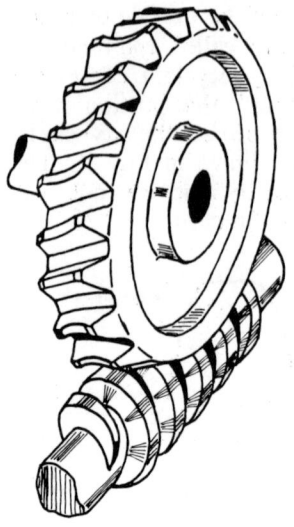

Fig. 36 Worm and worm-wheel

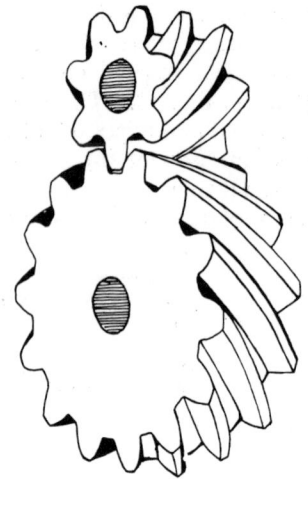

Fig. 37 Single helical gears connecting parallel shafts

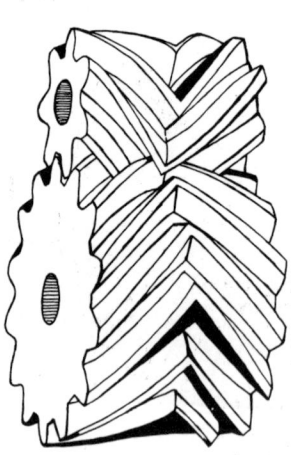

Fig. 38 Double helical gears

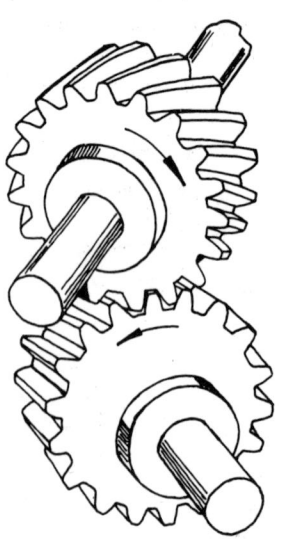

Fig. 39 Single helical gears connecting non-parallel shafts

they were produced. The angle between the two shafts can vary but is most commonly 90°. When shafts lie at right angles to each other but are not in the same plane, they can be connected by means of a worm and worm-wheel (Fig. 36). The worm is, in fact, a screw thread having an involute form and may have a number of starts. The worm engages in teeth cut into the worm-wheel.

Helical gears are similar to spur gears but instead of the teeth lying parallel to the axis of the gear they lie on a helix. By arranging the teeth in this manner, one pair of meshing teeth remains in contact until the following pair engages. A smoother drive is effected and the load transmitted is spread over more teeth. A single-helical gear drive (Fig. 37) produces considerable end thrust on the shafts and this must be catered for in the design of shaft housings. The end thrust can be eliminated if the gear is made in the form of a double helix (Fig. 38), i.e. one half of the face width of the gear is cut with a right-hand helix, the other half with a left-hand helix. Single helical gears, like worm gears, can also be used for shafts lying at right angles (Fig. 39).

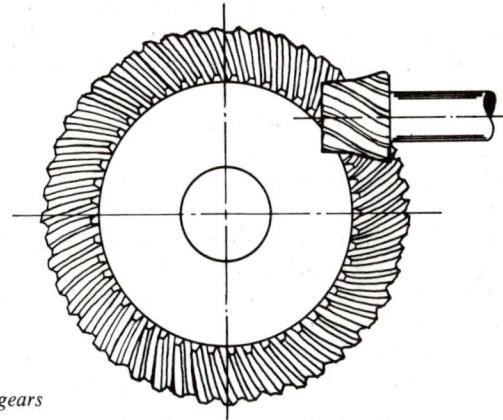

Fig. 40 Hypoid gears

Hypoid gears (Fig. 40) are often used in the automobile industry for the transmission of power from the propeller shaft to the differential gear. The shaft axes do not intersect and the gears are similar to bevel gears but the teeth are cut on *hyperboloids* instead of cones. (A hyperboloid is the solid generated by revolving a hyperbola about its axis.)

Our study of gears will be restricted to involute spur gears.

Involute spur gear terms

Fig. 41 shows the various parts and names of two gears in mesh. The explanation of each term is given below:

Pitch Circle: this circle represents the diameter of the plain friction cylinder. The diameter is the *pitch circle diameter*.

Base Circle: this is the circle upon which the involute curves are constructed.

Addendum Circle: this is outside diameter of the wheel. The blank gear wheel will be finish-machined to this diameter before the teeth are cut. Its diameter is often known as the *blank diameter*. This circle contains all the tops of the teeth.

Dedendum Circle: this circle contains all the bottoms of the tooth spaces. It is often known as the *root circle* and its diameter is the *root diameter*.

Addendum: this is the height of the tooth above the pitch circle.

Dedendum: this is the depth of the tooth below the pitch circle.

Whole Depth of a tooth is the sum of the addendum and the dedendum.

Clearance is the difference between addendum and dedendum.

Working Depth of a tooth is the maximum depth that one tooth extends into the mating tooth space. It is the sum of the addenda of the two gears.

Tooth Face: this is the working surface of a tooth above the pitch circle.

Tooth Flank: this is the working surface of a tooth below the pitch circle.

Circular Tooth Thickness is measured on the pitch circle and is the length of the arc across a tooth.

Chordal Tooth Thickness is the thickness of the tooth measured on the chord of the arc.

Circular Pitch is the distance from a point on one tooth to a similar point on the next tooth measured on the pitch circle.

Module: this is the pitch of gear teeth expressed as the amount of pitch circle diameter per tooth, e.g. if a gear has a pitch circle diameter of 50 mm and 25 teeth, the module is 2 mm.

When two gears are in mesh:

Centre Distance is the sum of the pitch circle radii. This distance can be varied by small amounts without affecting the function of the gears. If this distance is increased, more backlash in the gears will be experienced.

Pitch Point: this is the point of contact between the two pitch circles.

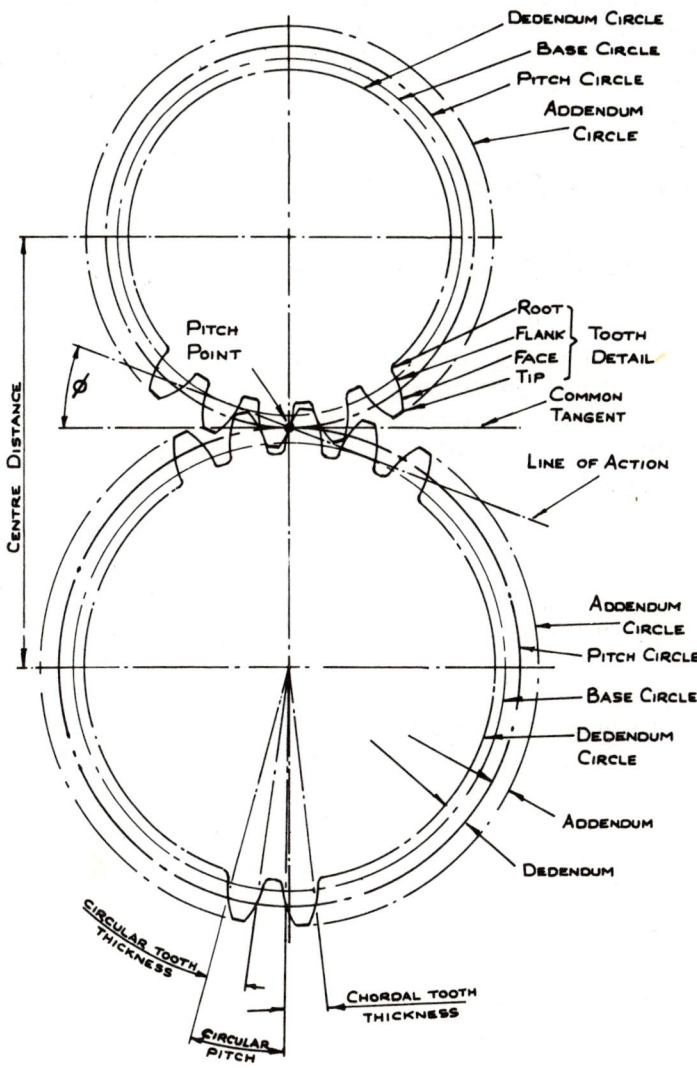

DEDENDUM CIRCLE
BASE CIRCLE
PITCH CIRCLE
ADDENDUM CIRCLE

PITCH POINT

ROOT
FLANK
FACE
TIP
TOOTH DETAIL

COMMON TANGENT

LINE OF ACTION

CENTRE DISTANCE

ø

ADDENDUM CIRCLE
PITCH CIRCLE
BASE CIRCLE
DEDENDUM CIRCLE
ADDENDUM
DEDENDUM

CIRCULAR TOOTH THICKNESS

CHORDAL TOOTH THICKNESS

CIRCULAR PITCH

Fig. 41 Involute spur gear terms

Line of Action: contact between the teeth of meshing gears takes place along a line which passes through the pitch point tangential to the two base circles; this is known as the line of action.

Pressure Angle: this is the angle between the line of action and a common tangent to the pitch circles passing through the pitch point.

Gear tooth proportions

The following abbreviations will be used:

M = module.

p = circular pitch.

N = number of teeth.

D = pitch circle diameter.

ψ = pressure angle.

Then:

$$\text{Module } M = \frac{\text{pitch circle dia.}}{\text{number of teeth}} = \frac{D}{N}$$

$$D = NM$$

$$\text{Circular pitch } p = \pi M = \pi \frac{D}{N}$$

$$\text{Circular tooth thickness} = \pi \frac{M}{2}$$

$$\text{Addendum} = M \qquad \text{Clearance} = 0{\cdot}25M$$

$$\text{Dedendum} = \text{addendum} + \text{clearance} = 1{\cdot}25M$$

$$\text{Blank diameter} = (N + 2)M$$

$$\text{Base circle diameter} = D.\cos \psi$$

The pressure angle of present day involute teeth is 20° (ISO Recommendations R53 and R54). This reduces the possibility of interference of the teeth in mesh and gives the teeth a wider and thus stronger root.

The width of the face of a gear depends upon the conditions under which it will have to work. For average conditions, a face width of 2–4 times the circular pitch will be satisfactory.

Involute gears will run correctly together if they have the same module, the same pressure angle and have been produced by the same manufacturing method.

Involute racks have the same proportions as the gears but the pitch circle diameter of the rack is infinitely large, i.e. it is a straight line. Racks have straight-sided teeth and the sides lie normal to the line of action. They are inclined to the vertical at an angle

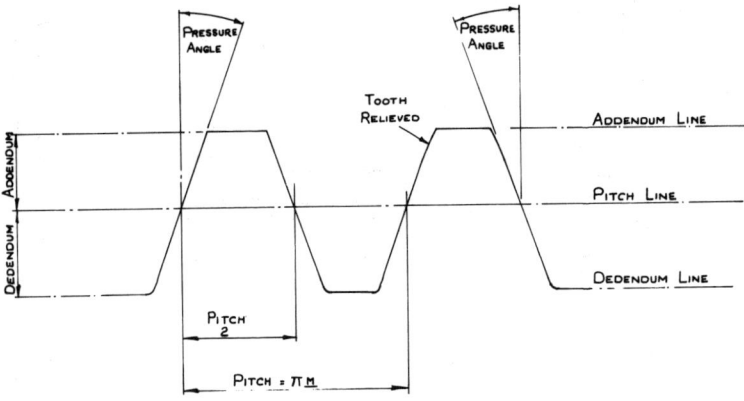

Fig. 42 Involute rack

equal to the pressure angle. In order to facilitate correct working of the rack in mesh with a gear, the tips of the rack teeth are slightly relieved (Fig. 42).

Conventional representation of gears and racks is shown in Fig. 43.

Example

A gear drive comprising two gears A and B is to have a ratio of 1·5:1. Gear A has 28 teeth, revolves at 126 rev/min and is the smaller of the two gears. If both gears have a module of 2 mm, determine:

    (*a*) the manufacturing particulars of each gear;
    (*b*) the speed of gear B;
    (*c*) the centre distance of the two gears.

71

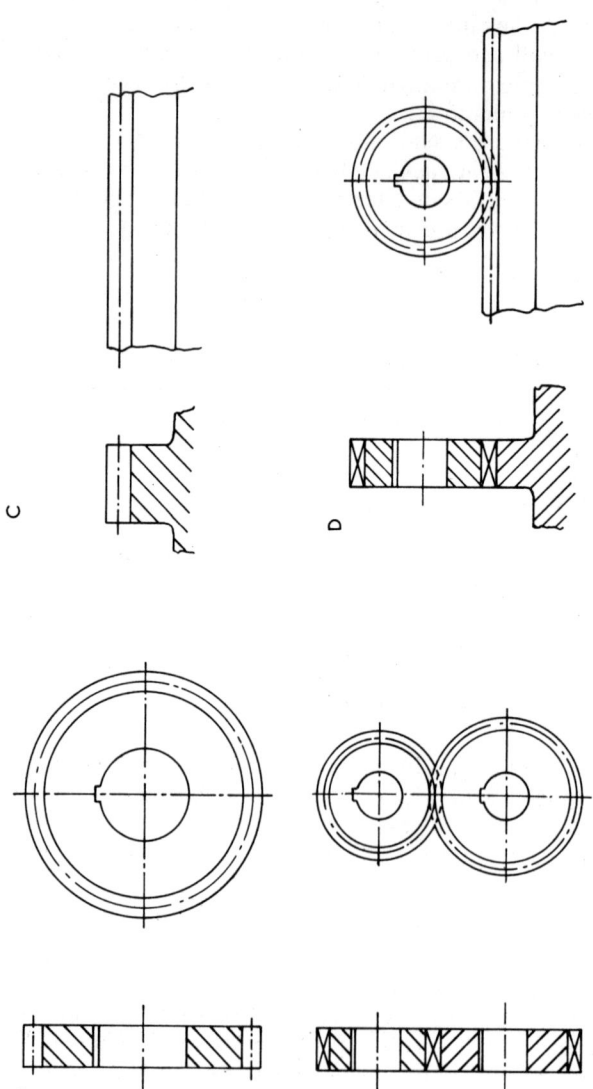

Fig. 43 *Conventional representation of gears A. Detail of spur gear. B. Detail of rack. C. Assembly of spur gear. D. Assembly of rack and spur gear*

Solution

(a)

Particular	Gear A	Gear B
Number of teeth N	28	$28 \times 1.5 = 42$
$D = NM$	$28 \times 2 = 56$ mm	$42 \times 2 = 84$ mm
Blank dia. $= (N+2)M$	$30 \times 2 = 60$ mm	$44 \times 2 = 88$ mm
Addendum $= M$	2 mm	2 mm
Dedendum $= 1.25M$	$1.25 \times 2 = 2.5$ mm	2.5 mm
Circular pitch $= \pi M$	$3.142 \times 2 = 6.284$ mm	6.284 mm
Circ. tooth thickness $= \dfrac{p}{2}$	$\dfrac{6.284}{2} = 3.142$ mm	3.142 mm

(b) The speeds of the gear wheels are inversely proportional to the number of their teeth,

i.e. $\quad \dfrac{\text{No. of teeth on A}}{\text{No. of teeth on B}} = \dfrac{\text{Speed of B}}{\text{Speed of A}}$

$\therefore \quad \dfrac{28}{42} = \dfrac{\text{Speed of B}}{126}$

$$\text{Speed of B} = \frac{28}{42} \times 126 = 84 \text{ rev/min.}$$

(c) Centre distance of mating gears = sum of pitch circle radii

$$= \frac{56}{2} + \frac{84}{2} = 28 + 42 = 70 \text{ mm}$$

EXERCISES

1. Calculate the manufacturing particulars of an involute spur gear having a module of 4 mm and 85 teeth with a pressure angle of 20°.

2. A 20 tooth pinion is in mesh with a rack. The teeth of the pinion and the rack have an addendum of 8 mm and a pressure angle of 20°. Sketch a few of the rack teeth stating on the sketch the critical manufacturing particulars. Tabulate the manufacturing particulars of the pinion.

3. The centre distance of two involute spur gears is 325 mm and the speed reduction is to be 2·25 : 1. If the module of the gears is to be 5, calculate:
   (*a*) the pitch circle diameters of the two gears;
   (*b*) the number of teeth on each gear;
   (*c*) the circular pitch of the teeth;
   (*d*) the addendum;
   (*e*) the dedendum.

4. A machine table is fitted with a rack having involute teeth with a pitch of 12·568 mm. This rack is operated by a pinion and an intermediate gear, the train giving a reduction of 4 : 1. If the pinion has a pitch circle diameter of 80 mm, calculate:
   (*a*) the module of the system;
   (*b*) the numbers of teeth on the pinion and the gears;
   (*c*) the distance moved by the table when the pinion makes a quarter of a revolution.

Answers

1. $D = 340$ mm; blank dia. $= 348$ mm; addendum $= 4$ mm; dedendum $= 5$ mm; circular pitch $= 12·568$ mm; circular tooth thickness $= 6·284$ mm.

2. $M = 8$ mm; dedendum $= 10$ mm; $p =$ pitch of rack $= 25·136$ mm; D of pinion $= 160$ mm; blank dia. $= 176$ mm.

3. (*a*) Pitch circle dias. $= 200$ mm and 450 mm
   (*b*) Nos. of teeth $= 40$ and 90
   (*c*) $p = 15·71$ mm
   (*d*) addendum $= 5$ mm
   (*e*) dedendum $= 6·25$ mm

4. (*a*) $M = 4$
   (*b*) Pinion $= 20$ teeth; gear $= 80$ teeth
   (*c*) 62·94 mm

CHAPTER 8

The Drawing of Accurate Profiles

Optical measuring devices of the profile-projection type include optical projectors and toolroom microscropes fitted with projection attachments. The silhouette image of a component with a critical profile is projected on to a screen and compared with an accurate enlarged drawing of the correct profile. Such components include form tools, form gauges, screw threads and gear teeth.

The projector enlarges the image of the profile to a scale depending upon the lens system. The magnifications which can be achieved vary according to the manufacture of the instrument but a survey of available instruments reveals magnifications of: 10 × ; 20 × ; 25 × ; 30 × ; 50 × ; 100 ×. This is not an exclusive list as magnifications other than these are available.

The drawing of the profile must be accurate, and for this reason, conventional drafting techniques are not employed. Tee square, set square and protractor are rejected and compasses or, better still, dividers, and straight edges only are used. For measuring distances, a steel rule is to be preferred and a table of tangents is utilized for setting out angles. Accurate work is essential. If an enlargement of 50 × magnification is being constructed, a drawing error of 1 mm would result in an error of 0·02 mm on the profile which would undoubtedly be critical. Errors of such magnitude are seldom made, but small drawing errors of up to ·05 mm can be tolerated since, at 50 × magnification, the resulting error in profile would be only some ·001 mm. Generally speaking, when a smaller magnification than 50 × is being used, large drawing errors can not be tolerated.

When making an accurate enlarged drawing, squareness of paper cannot be assumed, and the procedure is as follows:

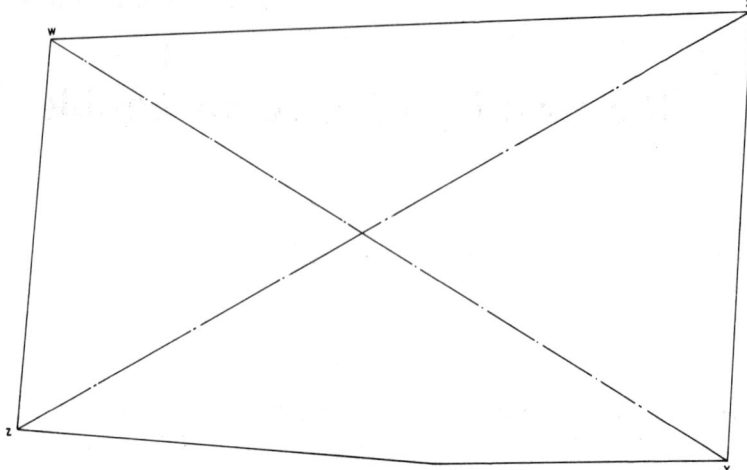

Fig. 44 Starting the layout

Fig. 44. The sheet of paper, WXYZ, has been fixed to the drawing board. The paper is obviously not very square and it is necessary to achieve squareness in order to proceed. Diagonal lines, WY and XZ, are drawn faintly on the paper.

Fig. 45. With the compass set at the largest practical radius and centre O, arcs are struck across the diagonals at A, B, C and D

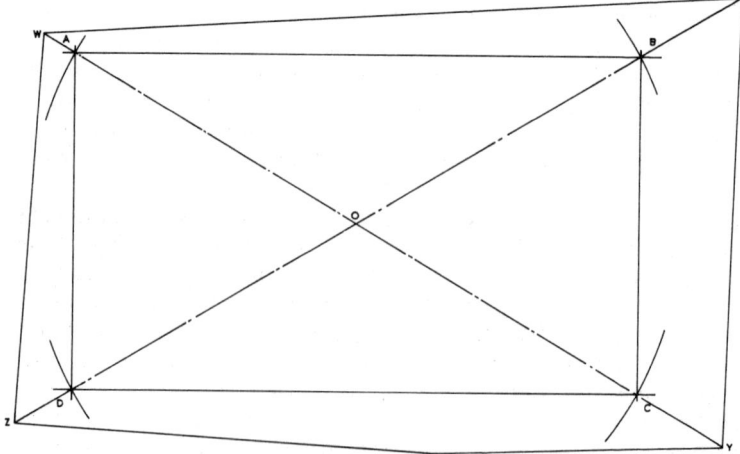

Fig. 45 Datum rectangle constructed

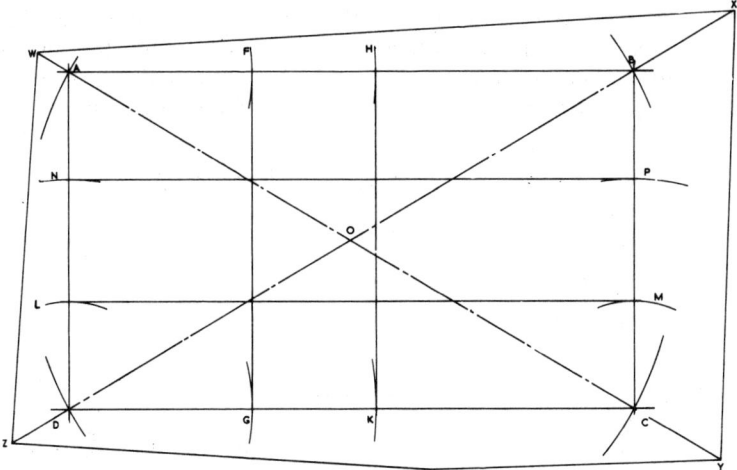

Fig. 46 Parallel and squared lines constructed

(beam compasses or trammells are useful here). When point A is joined to point B, point B joined to point C, point C joined to point D and point D joined to point A, a perfect rectangle has been constructed.

Fig. 46. Lines parallel to CD and AB can be constructed by setting the compass to any distance required and using points D and C as centres, arcs at L and M can be drawn. Line LM will be parallel to CD. A similar construction will produce line NP.

Lines FG and HK, parallel to AD and BC, can be constructed using a similar technique, but using points A and D as centres.

This basic rectangle is necessary to most profiles and it is from this that an accurate enlarged profile drawing is produced.

The use of a protractor for measuring angles is acceptable only when accuracy is not important. Enlarged profile drawings need to be accurate and the protractor should not be used for setting out angles. We have seen from Fig. 46 that it is possible to produce lines which are at right angles to each other. By marking off some predetermined distances along two adjacent lines at right angles, it is possible to obtain any desired angle. Fig. 47 shows two angles produced in this manner. The angle of 18° is produced by marking off distances of 10 units and 3·25 units on two lines at right angles. Now the tangent of 18° is 0·3249. This means that if the 'adjacent' side of this triangle was 1 unit, the length of the 'opposite' side would be 0·3249 units. As these values are small, and the unit is probably the centimetre, they are increased 10 times. If it is required to set out an angle of 30°, the tangent of the angle should be

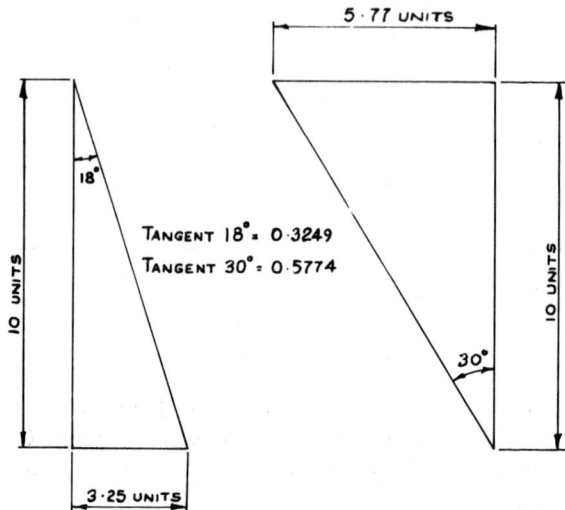

TANGENT 18° = 0·3249
TANGENT 30° = 0·5774

Fig. 47 Setting out the angles

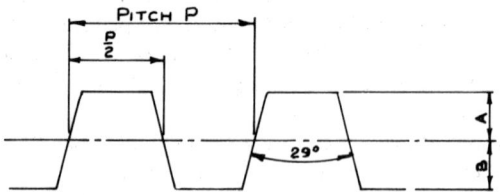

ACME THREAD DETAIL

5 mm PITCH

DIMENSION A = $\frac{P}{4}$

DIMENSION B = $\frac{P}{4}$ + 0·2mm

DRAW 1½ PITCHES TO A SUITABLY ENLARGED SCALE

Fig. 48 Acme thread detail

checked and found to be 0·5774. When this is multiplied by 10, it becomes 5·774 and when this value and 10 are set out on adjacent lines at right angles, an angle of 30° is constructed. It will be seen from Fig. 47 that 5·774 has become 5·77 and that 3·249 has become 3·25. This is done as it is impossible to measure to 0·01 mm with steel rule and compass or dividers.

Let us now apply the technique to an example. Fig. 48 shows the dimensional detail of an acme thread form whose profile is to be checked by profile projection. We are to draw $1\frac{1}{2}$ pitches of this thread to a suitably enlarged scale.

Assume that a magnification of 25 × will be suitable.
The dimensions of the thread are as follows:
Pitch of thread = 5 mm. When magnified 25 ×, this becomes 125 mm = P.

Half pitch $= \dfrac{5}{2} = 2\cdot5$ mm. When magnified 25 ×, this becomes 62·5 mm $= \dfrac{P}{2}$

$\dfrac{P}{4} = \dfrac{5}{4} = 1\cdot25$ mm. When magnified 25 ×, this becomes 31·25 = Dimension A.

$\dfrac{P}{4} + 0\cdot2$ mm $= \dfrac{5}{4} + 0\cdot2$ mm $= 1\cdot45$ mm. When magnified 25 ×,

this becomes 36·25 mm = Dimension B.

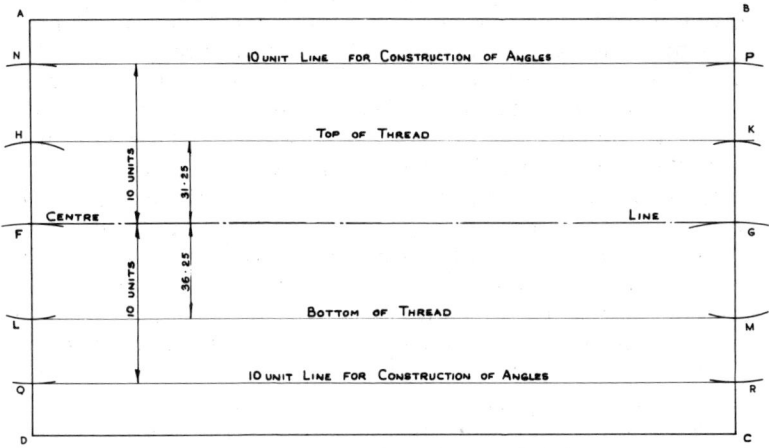

Fig. 49 Basic construction for acme thread

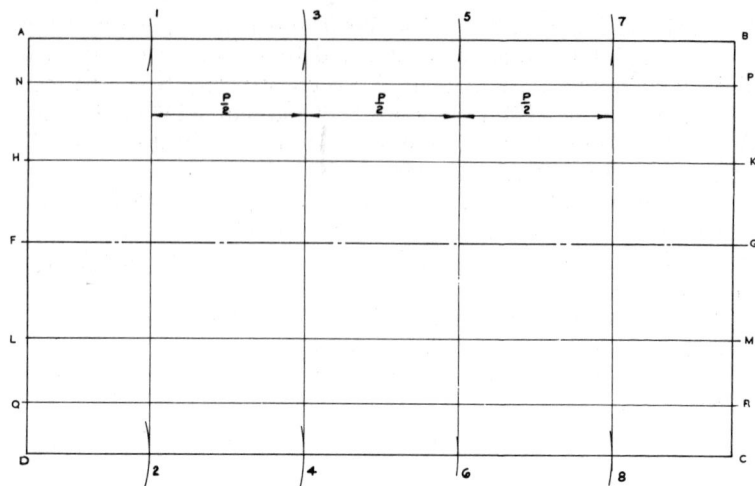

Fig. 50 Half pitch lines constructed for acme thread

Fig. 49 shows the basic true rectangle ABCD which has previously been drawn. As the thread form has a centre line which represents the effective diameter, a similar centre line can be constructed parallel to CD. The position of this centre line is not important. It should lie slightly nearer to AB than to CD. This centre line FG is constructed as previously outlined using points D and C as centres. It is now convenient to use this centre line as a datum for setting out the other horizontal lines. The line representing the top of the thread is next inserted by striking arcs of radius 31·25 mm from F and G. The line HK can then be drawn. Line LM, representing the bottom of the thread, is then constructed and is 36·25 mm from the centre line FG. Lines NP and QR are constructed 10 units above and below the centre line. These '10 unit' lines are used for the construction of the sloping sides of the thread form.

Fig. 50 shows all the previously constructed lines together with vertical lines representing the three half pitches which have to be drawn. The line 1,2 is positioned at any convenient point and from this are marked lines 3,4; 5,6; and 7,8. All these lines are parallel to AD and BC, and perpendicular to the previously constructed horizontal lines.

Fig. 51 shows the construction of the sloping lines. The sloping lines make an angle of $14\frac{1}{2}°$ with the vertical. The tangent of $14\frac{1}{2}°$

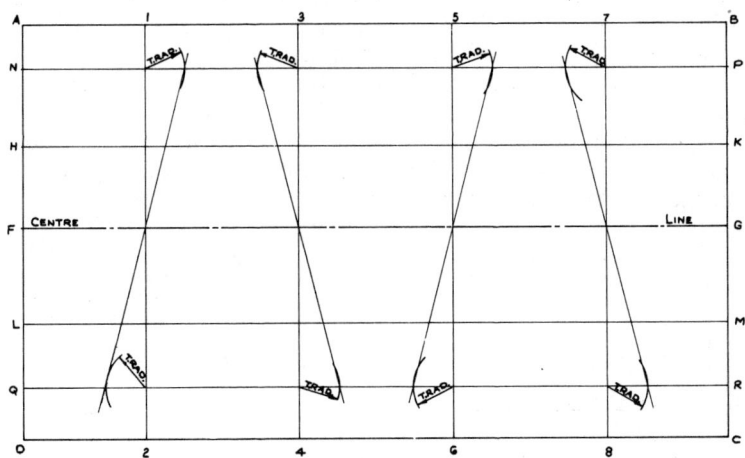

Fig. 51 Construction of angles for acme thread

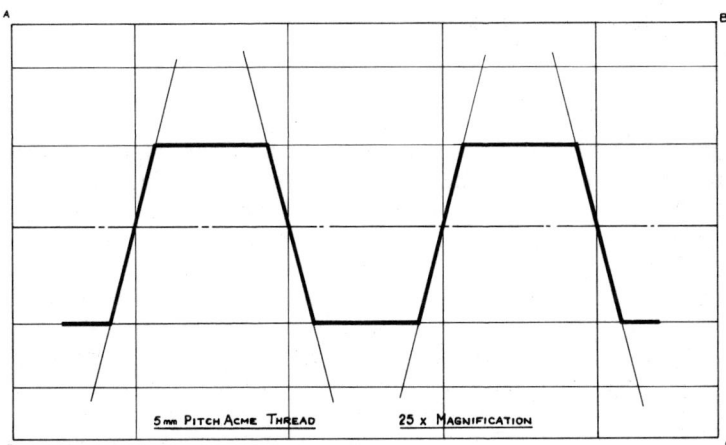

Fig. 52 The completed acme thread profile

is 0·2586. This when multiplied by 10 is 2·586 units. This value, shown as T radius on the drawing, is set off from the intersections of NP and QR with 1,2; 3,4; 5,6; and 7,8. At the points where the arcs cut NP and QR, the sloping lines should be drawn. The outline is complete but requires to have the line more clearly defined.

Fig. 52 shows the completed profile drawing with the constructional work behind it. The drawing is complete with a title and a statement of the magnification used.

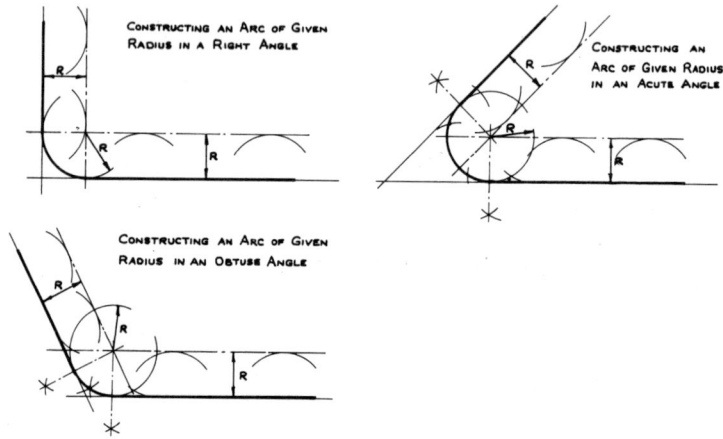

Fig. 53 Construction of arcs to straight lines

Many profiles have arc sections joined by straight lines or other arcs. Examination of the profile will reveal that the enlarged outline can initially be constructed of straight lines, the arcs being added to complete the outline. The constructions of the arcs follow the regular techniques of basic constructional geometry and these are given in Figs 53 and 54 for reference purposes. Fig. 53 shows the construction of arcs to straight lines. Fig. 54 shows the construction of arcs to other arcs. Notice that when a tangent to an arc is being constructed, the radius is constructed first so that it cuts the arc at the point where the tangent meets the curve.

The prohibitive cost of producing a component whose size and shape are perfect has led to engineers applying tolerance zones to those features of a component which are critical to its function. The production of a perfect profile is also costly and the provision of a suitable tolerance zone, within which the profile will be

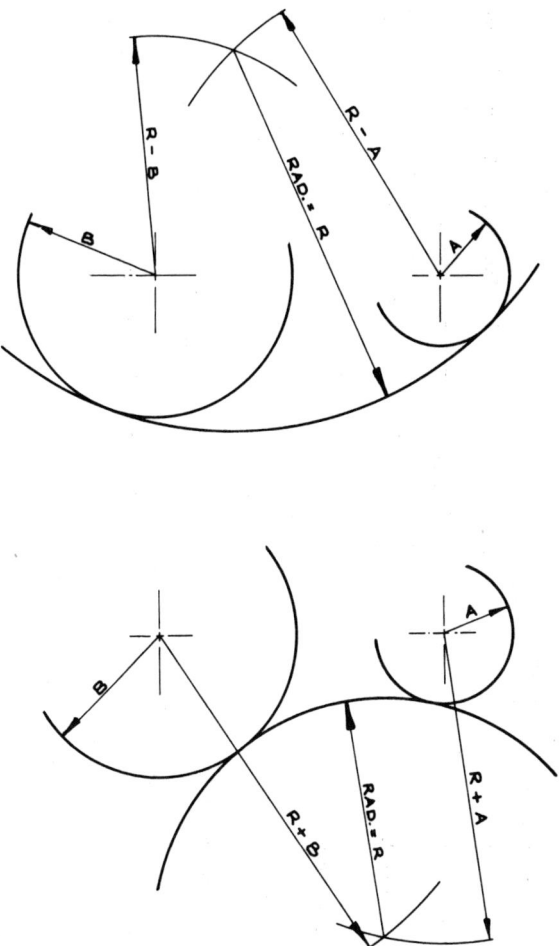

Fig. 54 Construction of arcs to arcs

acceptable, reduces the cost and allows for small errors in machine and operator performance. When a component's profile is checked by projection methods, the inspector should be able to check whether or not the profile lies within the tolerance zone specified by the designer. For this reason, the tolerance zone should be shown on the master outline. The tolerance zone can be unilateral,

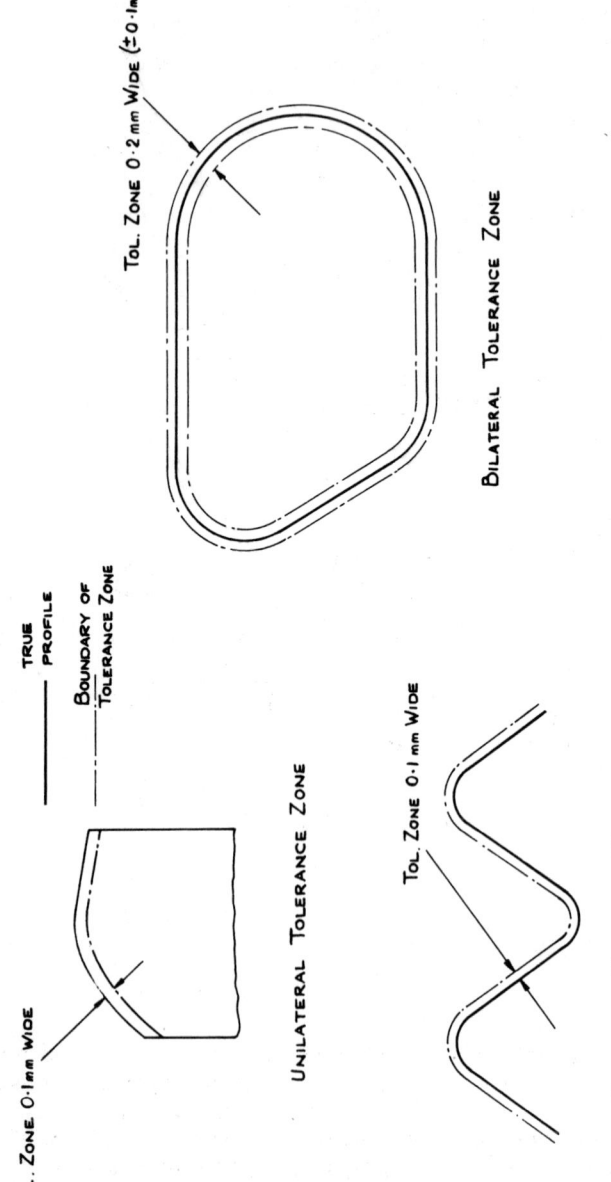

TOL. ZONE 0·1mm WIDE

TRUE PROFILE

BOUNDARY OF TOLERANCE ZONE

TOL. ZONE 0·2 mm WIDE (±0·1mm)

UNILATERAL TOLERANCE ZONE

TOL. ZONE 0·1 mm WIDE

UNILATERAL TOLERANCE ZONE

BILATERAL TOLERANCE ZONE

Fig. 55 Application of tolerance zones to accurate outlines

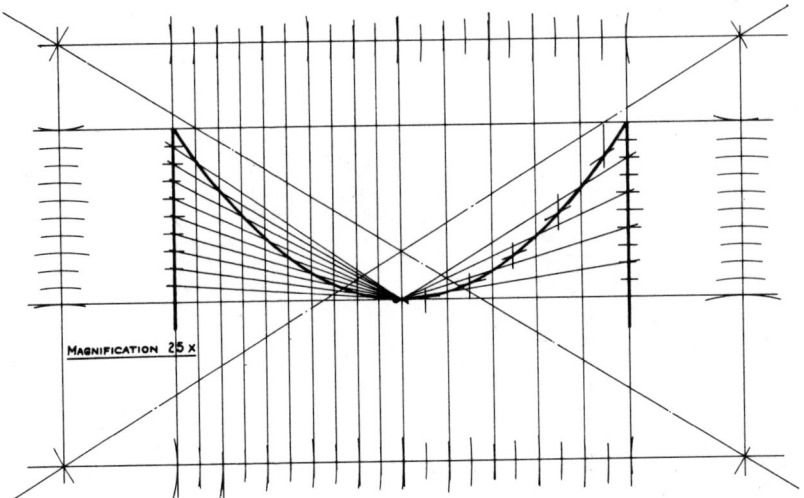

Fig. 56 Enlarged outline of shaving tool

i.e. displaced to one side of the true profile, or bilateral, i.e. displaced on both sides of the true profile. The selection of tolerance will depend upon the component's function. If the profile is that of an accurate hole used for measuring the flow of fluids, the tolerance zone may be placed bilaterally. If the profile is that of a screw thread then a unilateral tolerance zone will be applied to ensure adequate clearance in use. Fig. 55 shows the method of applying tolerance zones to accurate outlines.

Fig. 56 shows the construction, with the completed outline superimposed, of the accurate, enlarged drawing of the cutting edge of a shaving tool with a parabolic outline. The right hand side of the outline has been simplified having only part of the construction completed.

Fig. 57 shows the construction, tolerance zone and completed outline of a form tool for an internal form. The unilateral tolerance on the tool is arranged so that the form cut into the workpiece will not be less than the true profile.

When a permanent accurate, enlarged profile is required, the drawing is done on a material which is not affected by heat or moisture. This is to prevent any alteration of the profile drawing during storage. A typical material is one known as *astroscribe*.

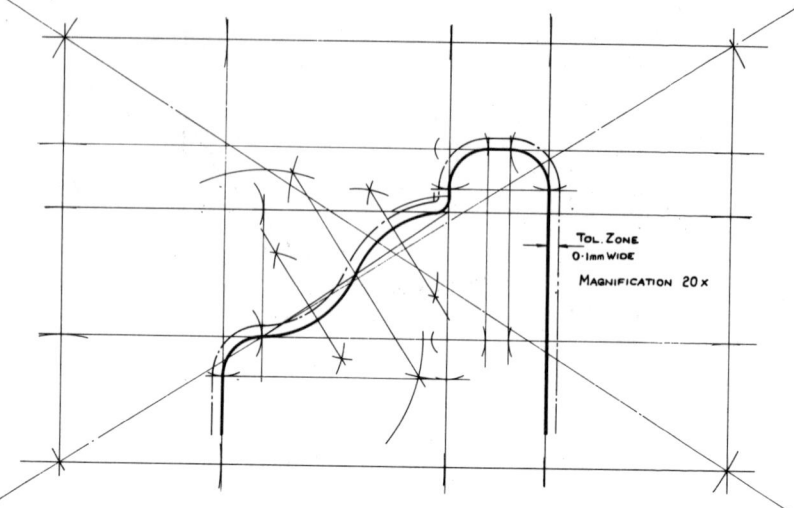

Fig. 57 Enlarged outline of form tool

This is an inert plastic material coated with a special wax on one side. The constructional work is actually cut into the wax with a scriber or stylus. When all the linework is completed, the actual profile lines are filled with ink. After the ink has dried, the wax which has been left on sheet is washed away with a solvent. This leaves the ink outline on a clear background, and all the unwanted constructional work is washed away with the wax.

EXERCISES

1. Draw accurately a master form for two teeth of a 2·5 mm module 20° pressure angle, involute rack which is to be checked by profile projection. Any normally available magnification can be used but this should be stated on the drawing. The master form should show a 0·1 mm tolerance at the tip and root of the teeth, the purpose of which is to ensure clearance in use.

2. Fig. 58 shows the proportions of a special screw thread applied to a male screw. Draw accurately, to a suitably enlarged scale, a master form for checking the form and pitch of such a screw having a pitch of 3 mm and a nominal outside diameter of 25 mm.

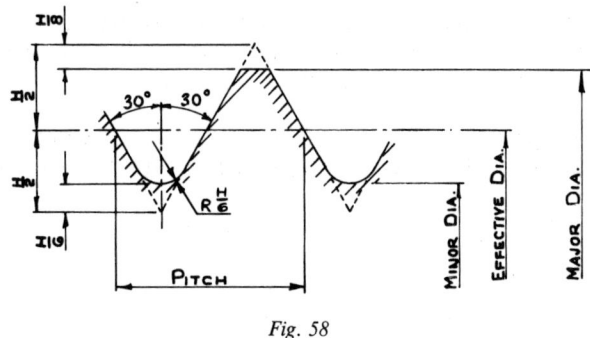

Fig. 58

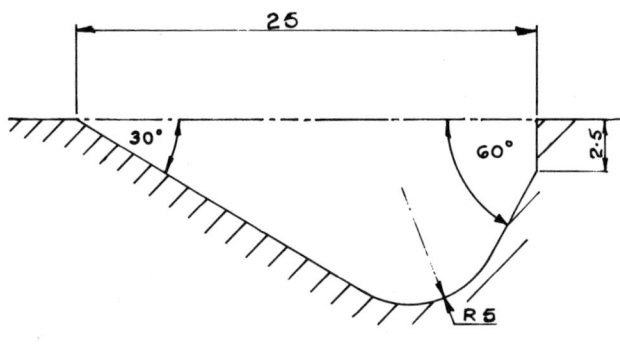

Fig. 59

3. The profile shown in Fig. 59 is to be cut into the periphery of circular shafts. Each shaft is to be checked by profile projection to find if the profile lies within the bilateral tolerance of $\pm\,0.1$ mm. Draw a master form, to any suitable magnification, showing the true profile and the tolerance zone.

4. By means of a number 1 gear tooth form milling cutter, machine four slots in a small piece of plate to represent three rack teeth of involute form. The pitch of the teeth can be obtained by indexing the milling machine table horizontally by means of the dividing head.

87

Produce a master form of three uncorrected involute rack teeth of the same diametral pitch and pressure angle to some suitable magnification. The master should be drawn on tracing paper.

On a profile projector, check the rack portion with the master. Comment on the pitch and form of the teeth. State any sources of error.

CHAPTER 9

Cutting Tools

The first consideration in the design and use of cutting tools is the actual selection of the cutting tool material. Prior to the year 1900 this was no problem because the only material available was a plain carbon steel ($1\cdot1$–$1\cdot3\%$ carbon), which had a limiting cutting speed of 6 m/min. It was limited to this speed because the tool would begin to soften or lose its temper at approximately 250°C. This restriction of cutting speed was a serious drawback to the development of production engineering only to be relieved by the development of high speed steel in the early part of this century, when it was found that additions of tungsten and chromium altered the cutting properties of tool steel. These additions allowed the tool to cut at temperatures of 550°C without losing its hardness, i.e. the property of red hardness was produced. A common high speed steel is designated 18 : 4 : 1, i.e. 18% tungsten, 4% chromium, 1% vanadium and possesses good general properties of toughness and red hardness. 18 : 4 : 2 has greater toughness but has a loss of red hardness. To impart greater red hardness properties the element cobalt is added; this produces a range of super high speed steel which requires care during heat-treatment because the cobalt has an embrittling effect which can crack the tool; consequently the cobalt addition must not exceed 12–13%. The common super high speed steels are 18 : 4 : 1 + 4% cobalt and 18 : 4 : 2 + 8% cobalt.

The next stage in the chase for greater red hardness was not a steel but a non-ferrous alloy of 45–50% cobalt: 30–35% chromium: 15–20% tungsten, known as *stellite*. It is cast to shape and derives its red hardness and cutting properties from the careful control of the casting process. When cold, stellite is as hard as high speed steel and retains its cutting properties up to 700°C, therefore no heat treatment is required. Compared with high speed steel it is expensive but permits higher speeds and feeds to be used giving a higher rate of output for the same tool life.

The advent of the sintered metals introduced another series of cutting tool materials known as the cemented carbides. The main cutting element is a carbide of tungsten with additions of other elements such as cobalt, which acts as a binder, and titanium and tantalum which reduce wear when cutting softer materials. The manufacture of these hard metals is referred to in Chapter 4, and due to the expense of having the material in a high purity powder form they prove to be costly in the initial purchase. They find their use in the form of tips which can be brazed or clamped to a steel holder, and are capable of high cutting speeds with a long life.

A recent entry into the field of cutting tool materials has been the ceramics. The term ceramic suggests the use of a clay in some form as the cutting material, but in fact the basis of this material is aluminium oxide which is also used as the abrasive in some grinding wheels. It is the method of manufacture which brings it, very broadly, into the field of ceramics, e.g. the fact that it is inorganic and formed at a temperature in excess of red heat. Again, due to cost and an inherent brittleness, it is used in the form of tips either bonded or clamped to a steel holder. These ceramic tips are chemically inert, do not weld to common metals, resist oxidation and retain strength up to 1000°C, giving rise to high cutting speeds with long tool life. The best results are obtained with ceramic tools at speeds in excess of those used for high speed steels and consequently this material has possibly outstripped the development of lathe gear speeds.

The last and most specialized of the cutting tool materials is the diamond. It has been used for some time in the field of fine finishing operations particularly precision boring, and more recently is finding application for turning the highly abrasive thermo-set plastics where it lasts substantially longer than other cutting tool materials.

A comparison in the cutting speeds obtainable from the major cutting tool materials when cutting mild steel can be seen in the following lists:

Plain carbon steel	12 m/min
High speed steel	30 m/min
Stellite	60 m/min

Cemented carbides   120 m/min
Ceramics   260 m/min

When selecting the best material each one must be considered in the light of the several different factors, i.e. the material being machined, whether intermittent cutting is being used, finish and accuracy required, condition of machine tools available and the production rate. It may well be that the most expensive in initial outlay may prove to be the most economical in the long term.

Despite the many efforts on all sides, the number of different types and shapes of lathe cutting tools used by industry is still very large. B.S. 1886 'Terms and Definitions for Single Point Cutting Tools', has cleared some of the confusion arising from nomenclature, and the various angles are fixed by virtue of both the tool material and the material being cut. Many existing tool shapes have undoubtedly emerged by virtue of the individual machinist experimenting to suit specific jobs.

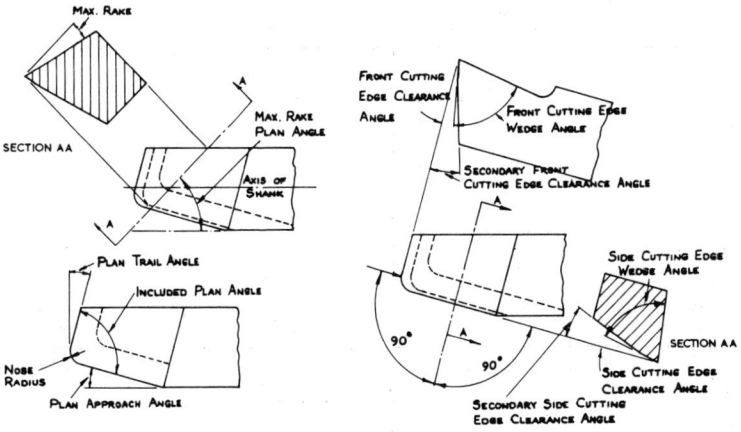

Fig. 60 Tool angles (from B.S. 1886: 1952, 'Terms and Definitions for Single Point Cutting Tools', reproduced by permission of the British Standards Institution, 2 Park St, London W.1, from whom copies of the complete standard may be obtained)

Fig. 60 shows an extract from B.S. 1886:1952 of the correct positions and names of the various angles applicable to lathe tools.

Fig. 61 is a typical manufacturers tool shape chart, in this case issued by Hall and Pickles, Ltd, from which the machinist can select the tools he requires. It will be appreciated that the ordering

Light Turning and Facing	Straight Nosed Rougher	Curved Nose Rougher	Knife or Side Cutting
T.R.16° SHAPE No. 1	T.R.16° 3	T.R 16° 5	T.R.14° 7
Specify No. 2 if Left Hand Tool is required	Opposite hand No 4	Opposite hand No. 6	Opposite hand No. 8
Down Cutting and Facing	Parting off	Round Nosed Planer or Shaper	Stub Nosed Planer or Shaper
T.R.12° 14	T.R.5° 16	T R 12° 17	T.R.12 18
Opposite hand No. 15	If blade preferred offset on other side. Specify 16 L.H.	Cuts in either direction	Cuts in either direction
Right Angle Parting off	Square Nosed Turning and Facing	Broad Nosed Facing	Turning and Facing for Boring Mills. Auto Combination and Capstan Lathes
T R 5° 27	45° T R 12° 29	T.R.12° 31	T.R.7° 33
Opposite Hand No. 28	Opposite hand No 30	Opposite hand No. 32	Opposite hand No. 34
Diamond Nose	Light Turning and Facing in awkward corners	Hardened Blank for grinding into form and radius tools at customer's works	Bar Boring
T.R.12° 43	T.R.16° 45	NO TOP RAKE 47	T R 11° 48
Cuts in either direction	Opposite hand No. 46	Also suitable for tools for work on brass and gunmetal	Opposite hand No. 49

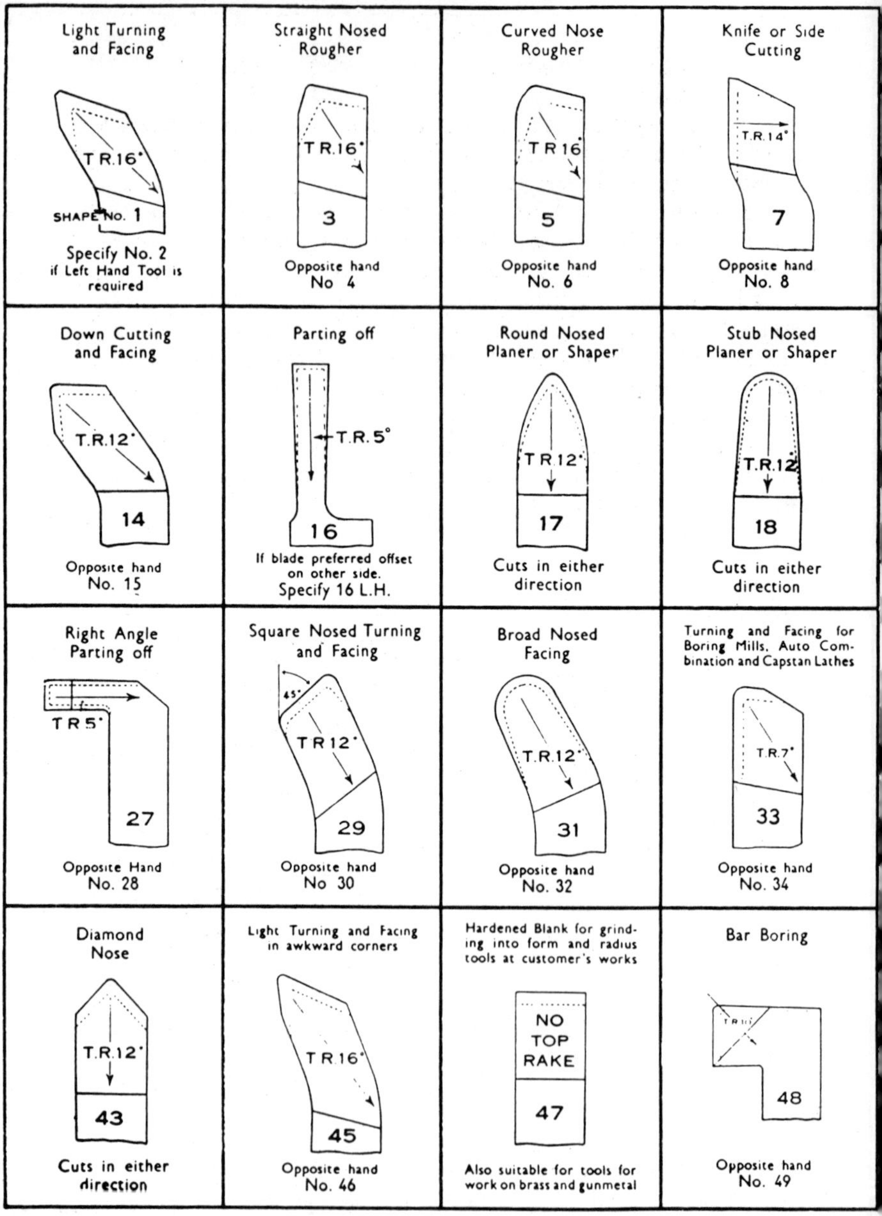

Fig. 61 Shape chart of hydraweled tools manufactured by Hall and Pickles, Ltd.

Bar Turning	Plain Form	Finishing	External Screw Cutting
T.R.20° **9** Opposite hand No. 10	T.R.12° **11** Cuts R.H. or L.H.	T.R 12° **12** Cuts R.H. or L.H	T R 10° **13** If required to cut L H. threads, specify 13 L.H.
Facing	Round Nosed Rougher	Heavy Duty Turning	Right Angle Recessing
T.R.16° **19** Opposite hand No. 20	T.R.16° **21** Opposite hand No. 22	45° T.R.16° **23** Opposite hand No. 24	T.R.5° **25** Opposite hand No. 26
Light Turning and Facing for Boring Mills, Auto Combination and Capstan Lathes	Straight Rougher for Manganese and other Hard Tough Steels	Crank Turning, Recessing, and Finish internal Boring	Mild Steel Turning and Facing for Boring Mills, Auto Combination and Capstan Lathes
T.R.7° **35** Opposite hand No. 36	30° T.R.12° **37** Opposite hand No. 38	T.R 12° **39** Opposite hand No. 40	T R 20° **41** Opposite hand No. 42

Boring	Swan Necked Finisher	Swan Necked Rougher
T R 5° **50** 50A Vee Nose for internal screw cutting 50B Round Nose	**52** T.R 12° Cuts in either direction	**53** T R 12° Cuts in either direction

of tools is much facilitated by reference to the charts since the general ordering procedure follows the pattern of machinist–foremen–purchasing officer–supplier.

It will be noted in No. 2 that T.R. is in fact true rake, which corresponds to the line of greatest slope. When only a back rake or a side rake angle is used, that angle will also be the true rake angle. However, where both side and back rake angles are employed on the same tool, the true rake is the combination of these two angles. It is in fact, the direction in which the chips will flow down the tool face.

On examination of the tool shapes available it is noticeable that tools designated as roughers or heavy duty have either a pronounced plan approach angle or a curved approach cutting edge. The reason for this design is that for a given depth of cut, the cutting load is spread along a longer face than if a standard knife tool, i.e. no plan approach angle, was used.

The tools numbered 52 and 53 on Fig. 61, termed swan necked rougher and finisher, deserve some mention because of their unusual shape. The purpose of this lies in the fact that if the tool meets a hard spot in the work, the cutting edge is forced down and away from the work because it is below the support shoulder of the tool post. With conventional tools the cutting edge is actually above this support shoulder with the result that the cutting edge is forced further into the work ruining the finish. The swan necked tool finds use in form work and also as a shaper tool where the conditions outlined above still apply.

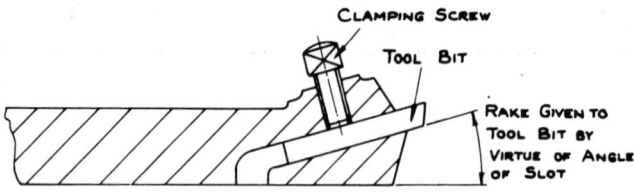

Fig. 62 Typical holder for square tool bits

Fig. 62 shows a tool-holder with the method of holding the tool bit in position. The bit is usually made up entirely of the cutting material but sometimes it may have a tip of diamond or special cutting alloy. The bits are generally supplied unformed so that the tool grinder can produce the angles appropriate to the job required.

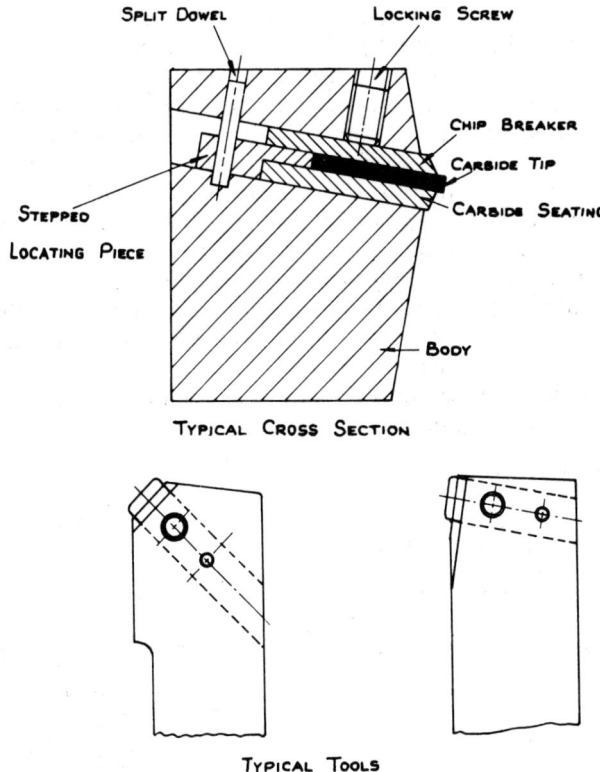

SPLIT DOWEL   LOCKING SCREW

CHIP BREAKER

CARBIDE TIP

STEPPED

LOCATING PIECE

CARBIDE SEATING

BODY

TYPICAL CROSS SECTION

TYPICAL TOOLS

Fig. 63 Tools with throw-away carbide tips

Fig. 63 shows a rapidly developing form of lathe tool-holder for what is known as *throw-away tip machining*. The manufacturers of cemented-carbide tips have developed them so that it is cheaper to provide a new tip than to regrind the worn one. The tip shown can be obtained in either a triangular or a square shape to suit the particular holder so that, as one cutting edge loses its sharpness, the tip can be indexed by slackening the screw to present a new cutting edge to the work. Locating faces ensure that the tool bit is always in the same position. On further inspection of Fig. 63, it will be noticed that the cutting rake angle is, in fact, negative when compared with the conventional rake angle associated with the tool angles designated in B.S. 1886 (see Fig. 60). The use of this negative rake cutting has been introduced to counteract the relative brittleness of the carbide tips by causing the chips produced to contact the tool at a point further back from the cutting edge. The forces involved in the cutting are then absorbed by the support of the tool-holder instead of by the tool bit. However, throw-away tip tool-holders can be obtained with positive rakes for use with soft or free cutting materials; here the cutting forces will be relatively light. With positive rake cutting clearance has to be provided on the tip so that only the top edges can be used, whereas with negative rake cutting the position of the bit provides the clearance giving double the cutting edges, i.e. top and bottom. This throw-away principle has been so successful that it has been extended to inserted teeth for milling cutters.

Fig. 64 shows three types of form tool commonly used in both centre and capstan lathe work.

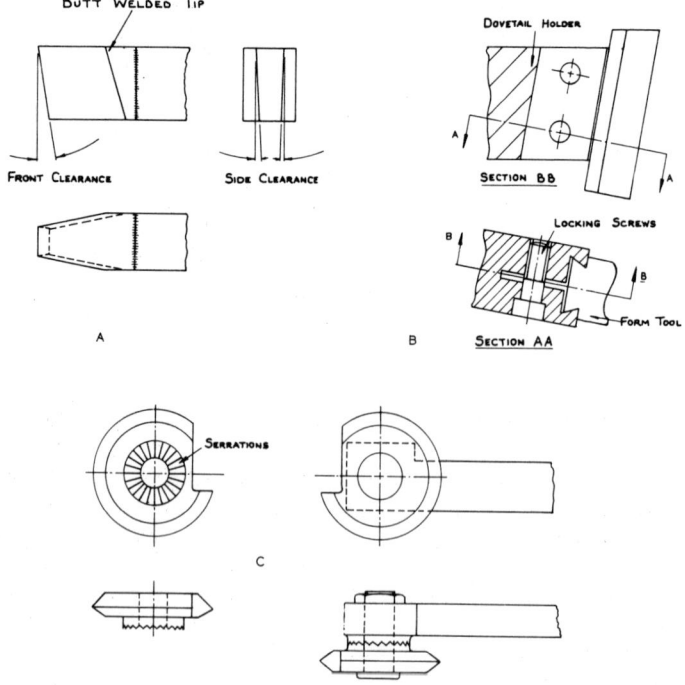

Fig. 64 Form tools A. Flat form tool. B. Dovetail form tool. C. Circular form tool and circular form tool in holder

(A) *Flat form tool:* with this type, a standard shank is used and a cutting alloy is butt-welded to it. This simple type of form is used to provide an inexpensive form tool with a short life.

(B) *Dovetail form tool:* consists of a block of cutting material with the form ground on the full length of the front face. On the back face is a dovetail to fit a special holder which is generally found on capstan or turret lathes. As the cutting edge wears, only the top face is ground and the form is retained until the block is too small to fit the holder. The tool is limited to narrow forms because of the lack of rigidity caused by its overhang.

(C) *Circular form tool:* the required form is turned or ground on a circular blank of the cutting alloy and a cutting edge is provided by removing a segment as shown. As the cutting edge wears, the tool is indexed one serration and the top surface is ground to the correct position. This tool also is limited to narrow forms because of the lack of rigidity caused by its overhang.

EXERCISES

1. Sketch two views of the finished tool, showing all the angles, from the following information:

   Shank 30 mm × 25 mm × 200 mm
   Tip   22 mm × 16 mm × 8 mm

   Tool nose angles: back rake 0°; side rake 5° negative.
   Side and Front clearance angles 5°; plan approach angles 10°; plan trailing angle 20°.

   (C.G.L.I.)

2. Make a neat sketch of a roughing tool suitable for turning a steel casting which involves intermittent cutting. Show clearly the relevant tool angles and state the material from which the tool is to be made.

3. Describe the action of a chip-breaker and give a sketch of three different types of chip-breaker used on lathe tools.

4. Show by means of a sketch, any 'throw-away tip' tool-holder with which you are familiar, emphasizing the tool-location, tool angles and locking device.

5. Discuss the importance of back rake when applied to the design of form tools.

6. Discuss the advantages and disadvantages of circular and flat form tools for producing standard forms such as radii and angles on a lathe. Sketch a tool-holder suitable for holding circular form tools.

CHAPTER 10

Gauge Design

Gauges are used to inspect components or certain features of components and although no actual measured reading is obtained, it can be ensured that the dimension being inspected is within the designer's limits of size. The *'Go' gauge* checks the maximum metal condition in that it has to pass over or through a particular feature. Conversely the *'Not Go' gauge* checks the minimum metal condition and consequently it should not pass over or through a feature. Fig. 65 illustrates the terms of maximum and minimum metal conditions. It can be appreciated that in normal centre lathe turning for one-off or small batch production, the operator will work to the maximum metal limit in that he will adjust the depths of cut so that

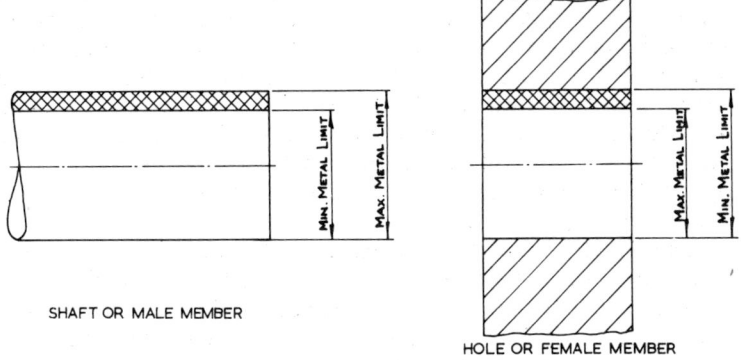

SHAFT OR MALE MEMBER

HOLE OR FEMALE MEMBER

Fig. 65 Maximum and minimum metal limits

the finished shaft or bore is just inside the maximum metal limit. However, in setting the tools for automatic or capstan lathes in mass production work, the setter will set the tools to cut initially on the minimum metal limit. Then as the tool wears, the components produced will progressively approach the maximum metal limits, i.e. shafts will become larger whilst bores will become smaller. The whole of the tolerance permitted by the designer will then have been utilized. By sample inspection the setter can predict when the work will be outside the allotted limits necessitating a reset of the tools.

There are two British Standard Specifications available for gauge design: B.S.S. 1044: 1942 'Recommended Designs for Gauges' and B.S.S. 969: 1953 'Plain Limit Gauges: Limits and Tolerances'.

Taylor's Principle

This states that the 'Go' *gauge* should be of 'full form' in order to inspect the maximum metal limits of as many dimensions as possible in one gauging operation, whilst the 'Not Go' gauge should inspect the minimum metal condition of only one feature or dimension at a time.

Applying this principle to the gauging of a bore we find that the 'Go' gauge should be a cylindrical plug, i.e. a full circular section, and of the same length as the bore. This controls the diameter at any given point whilst at the same time ensuring parallelism and roundness throughout the bore. In practice however, it is accepted that the method of manufacture can, in most cases, control the features of parallelism and roundness which means the plug gauge can be of a more convenient length. The optimum 'Not Go' gauge should be of a plate or pin form in order to inspect one diameter at a time. It will be appreciated that the normal 'Not Go' plug gauge is of full form thereby accepting a bore which has some diameters outside the specified limits of size. Fig. 66, illustrates the point in question so that in the square hole which has been accepted by the 'Not Go' gauge, one of the dimensions is outside the tolerance zone.

Applying the principle to the gauging of shafts, the only acceptable 'Go' gauge would be a cylindrical ring to give the full form requirement, whilst a gap or snap gauge would provide the 'Not Go' requirement. In practice the latter type of gauge carries out both the 'Go' and 'Not Go' inspections, and leaves the inspection of roundness etc. to other types of equipment.

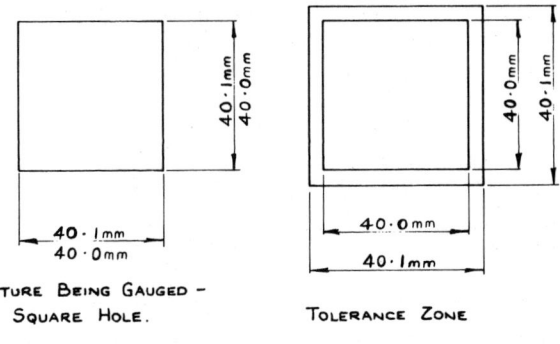

FEATURE BEING GAUGED –
SQUARE HOLE.

TOLERANCE ZONE

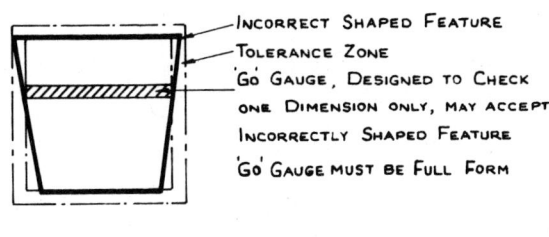

INCORRECT SHAPED FEATURE

TOLERANCE ZONE

'GO' GAUGE, DESIGNED TO CHECK
ONE DIMENSION ONLY, MAY ACCEPT
INCORRECTLY SHAPED FEATURE

'GO' GAUGE MUST BE FULL FORM

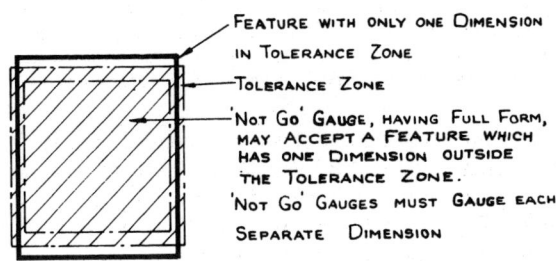

FEATURE WITH ONLY ONE DIMENSION
IN TOLERANCE ZONE

TOLERANCE ZONE

'NOT GO' GAUGE, HAVING FULL FORM,
MAY ACCEPT A FEATURE WHICH
HAS ONE DIMENSION OUTSIDE
THE TOLERANCE ZONE.

'NOT GO' GAUGES MUST GAUGE EACH
SEPARATE DIMENSION

Fig. 66 Taylor's Principle

Plug Gauges

Fig. 67 illustrates variations in the type of plain plug gauge available. The 'Go' end is of the required full form and a convenient length, following Taylor's Principle where possible. The 'Not Go' end is of full form for convenience of manufacture, with the length

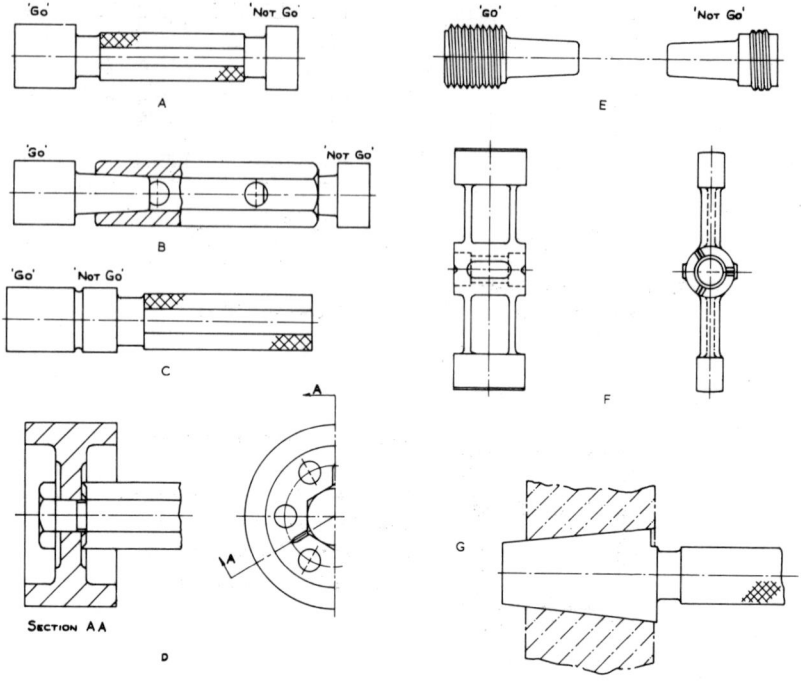

Fig. 67 Plug gauges A. Double ended plain plug gauge. B. Double ended renewable end type with taper lock handle. C. Single end progressive plug gauge. D. Plain plug gauge with tri-lock handle. E. Screw plug gauges for taper lock handle. F. Bar gauge for use with tri-lock handle. G. Taper plug gauge.

kept to a minimum because it is not intended that it should enter the bore. Generally, the 'Go' and 'Not Go' members are combined on one gauge to form a double ended or a progressive type of plain plug gauge. The gauging members can be integral with the handle or they can be separate so that one holder can serve several gauging members, as illustrated at B.

For sizes greater than 54 mm and up to 200 mm diameter, the tri-lock method of attaching the gauging member to a holder is recommended. This is illustrated in Fig. 67D; the gauge handle has three V-shaped projections at 120° to each other whilst the gauging member has three corresponding slots at 120°. On assembly the projections and slots are mated, with the handle and gauging member held together by a central locking screw. Due to the size of these gauges, some metal has to be removed to reduce the weight and thereby reduce operator fatigue; this is accomplished by recessing the member, front and rear, which also accommodates the locking arrangement. An additional weight reduction is achieved by drilling a series of holes inside the recesses, which also serve to release any trapped air when inspecting a blind hole.

When inspecting bores of more than 115 mm diameter, a full form gauge would be cumbersome and prove inconvenient for practical purposes, therefore a part form gauge known as a bar gauge is used (Fig. 67F). It will be appreciated that for a 'Go' gauge this does not satisfy the conditions previously laid down, but in this instance, practical considerations take precedence.

In the case of inspecting internal screw threads, a full form 'Go' gauge is used with a truncated short thread for the 'Not Go' member. The 'Go' ensures that the thread is not too tight whilst the 'Not Go' will reject a slack thread. These are illustrated in Fig. 67E.

A plug gauge for the inspection of tapered holes is shown in Fig. 67G. The limits of size are represented by the step shown at the larger diameter. If the hole is correct, the top of the hole should lie in the limits of the step when the plug has been inserted as far as it will go.

In the manufacture of plug gauges the centres used should be of good quality with the length of the cone kept short. Also, the centres should be protected by means of a recess. For gauges in excess of 65 mm diameter, an extension is provided to prevent the possibility of burring (see Fig. 68).

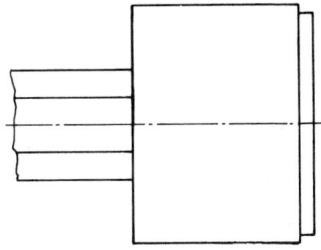

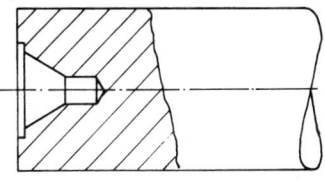

GAUGE EXTENSION
USED TO PROTECT THE ENDS OF
GAUGES OVER Ø 65 mm AGAINST
ACCIDENTAL BURRING.

PROTECTED CENTRE
GAUGE ENDS ARE RECESSED BY ·75mm TO
1·5 mm TO PROTECT THE CENTRE CONE.

Fig. 68 Plug gauge details

Fig. 69 shows various ring gauges in current use. It should be noted that the taper ring gauge, shown at C, has a step machined on one surface and the inspection of male tapers is carried out in a similar manner to the inspection outlined for female tapers.

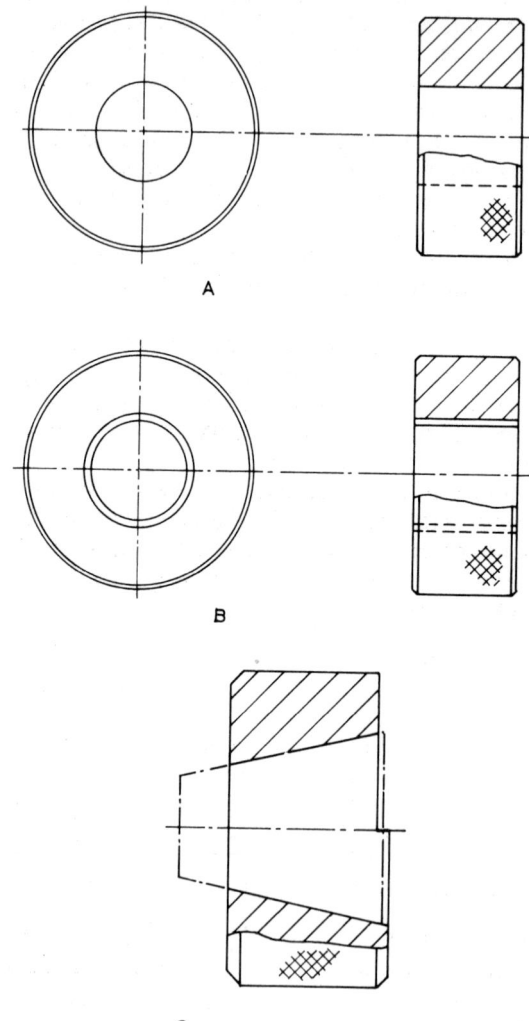

Fig. 69 Ring gauges A. Plain ring gauge. B. Screwed ring gauge. C. Taper ring gauge

Fig. 70 shows a variety of gap gauges. The advantages of the adjustable type are that, within limits, one gauge will cover a range of sizes and also wear can be taken up. Normally the gauges are set in the inspection department and the adjusting screws are

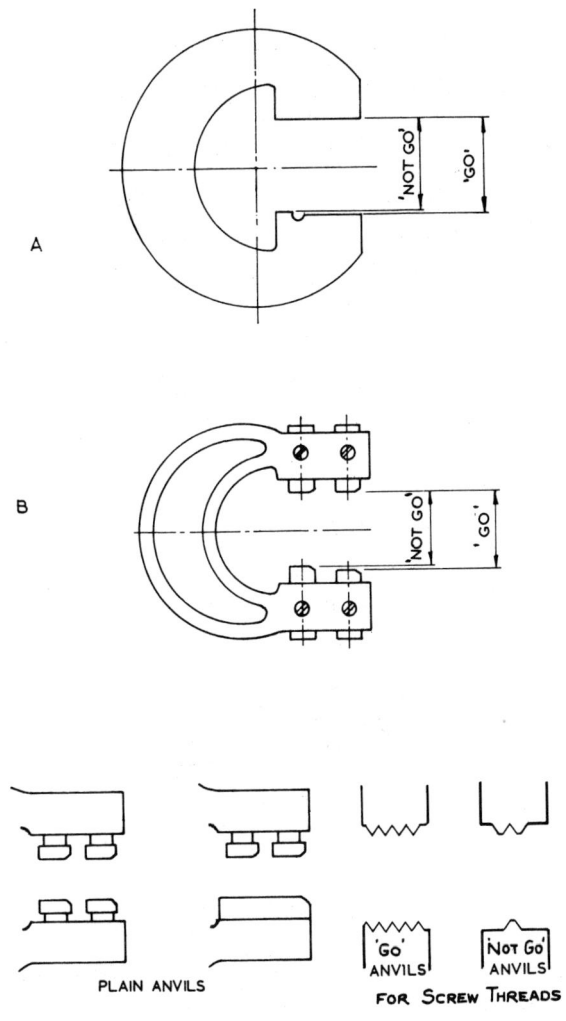

Fig. 70 Gap gauges A. Plain gap gauge. B. Adjustable gap gauge. C. Alternative anvils for adjustable gap gauges

sealed and stamped to prevent unauthorized alteration during a production run. Insulated finger grips are sometimes fitted to prevent the heat from the hand causing expansion of the gauge during prolonged usage.

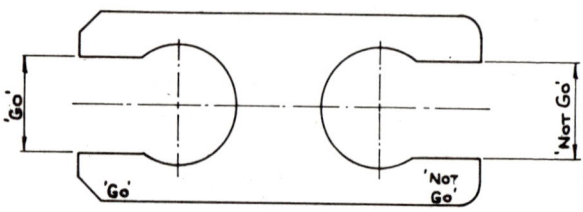

Fig. 71 Double ended snap gauge

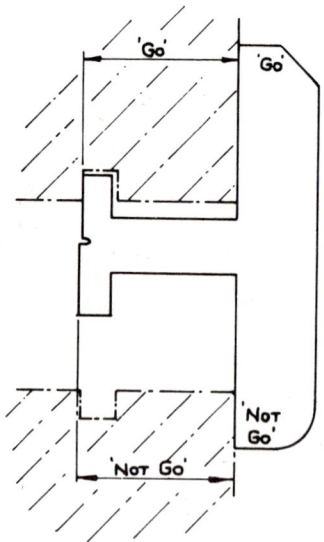

Fig. 72 Recess gauge

Fig. 71 shows a double-ended snap gauge which is used mainly on small pins and shafts.

Fig. 72 shows a special plate gauge designed for the inspection of the recess. The 'Go' and 'Not Go' members are both taken from a common datum, i.e. the underside of the head.

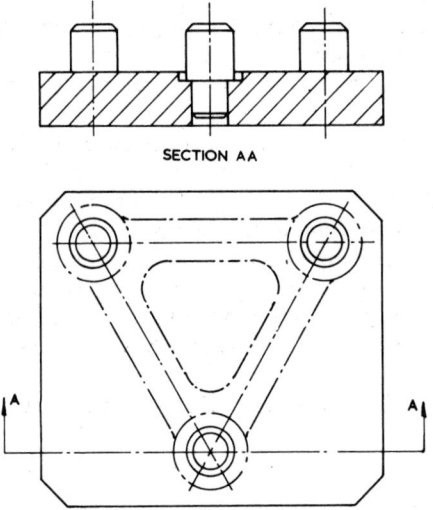

Fig. 73 Position gauge

Fig. 73 shows a form of·position gauge. This type of gauge inspects the relative position of several features either to each other or to a fixed datum. The position gauge illustrated is inspecting the position of the three holes in relationship to each other.

Fig. 74 shows another special type of gauge known as a *receiver gauge*. It is, in fact, an inspection fixture which can be made to inspect several features at the same time. The feature to be inspected in the example is the centre height of the bored boss. The pin of

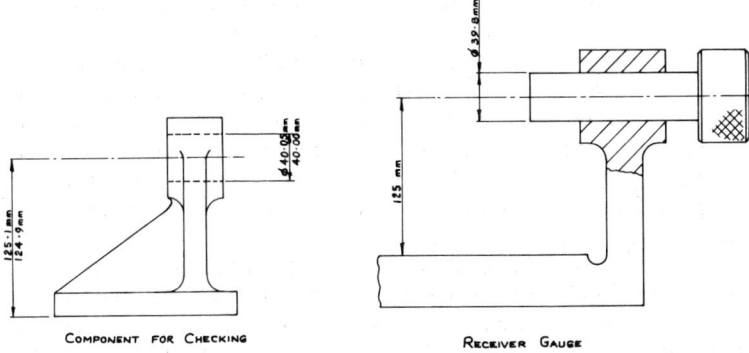

COMPONENT FOR CHECKING

RECEIVER GAUGE

Fig. 74 Receiver gauge

107

the receiver gauge has to have a dimension such that it will enter components whose centre height is inside the limits but will not enter those with a centre height outside the limits. Consider a component with a centre height on the top limit of 125·1 mm then the bottom of the bore would be a distance of 125·1 mm – 20 mm or 105·1 mm above the base. The corresponding bottom of the pin would be 125 mm – 19·9 mm or 105·1 mm above the base, and therefore the component would be accepted.

Similarly for a component with a centre height on the bottom limit of 124·9 mm then the top of the bore would be a distance of 124·9 mm + 20 mm or 144·9 mm above the base. The corresponding top of the pin would be 144·9 mm above the base and again the component is accepted.

The bore itself would be inspected separately by a standard limit plug gauge.

Gauge Materials

A variety of materials is available for the manufacture of gauges, examples of which are:

(*a*) *Mild steel* (*case hardened*). This low carbon steel is used for the majority of small- and medium-sized gauges and gauging fixtures. It has the advantages of good machinability and stability, and it can be surface hardened to varying depths at any required position.

(*b*) *Gauge Plate.* This is a 0·7–1·1% carbon steel for either oil or water quenching. Oil quenching is preferable to water quenching since there is less chance of cracking. Stock materials such as gauge plate and flat ground stock fall into this category, and they may be used in the soft or hardened condition.

(*c*) *Cast Iron.* Good quality close grained cast iron is used in the bodies of larger gauges and fixtures. Ageing of the casting is most important and this can be done either artificially, by heat treatment, or naturally; the latter method being preferred. It consists of rough machining the casting and then leaving it to weather in the work-shop yard for about six months. In this way the stresses set up in the casting process are removed.

(*d*) *Deposited Materials.* Deposition of hard metals, particularly chromium, by electro-plating directly on to steel has been developed and is now widely used. The most accurate method is to machine the gauge under size, plate a few thousandths of an inch oversize and then finally grind and lap to size.

(*e*) *Tungsten Carbide*. This material is being increasingly used for tipping the anvils of micrometers, indicators, and other reference surfaces which are subject to constant abrasion.

EXERCISES

1. (*a*) Discuss the advantages of using adjustable type over plain, gap gauges.

   (*b*) Explain with the aid of sketches, one method of adjusting the anvils and a method of locking them.

2. A 45 mm diameter N8 bore is to be checked with a limit plug gauge. Sketch a suitable gauge for this operation and state the tolerances allowed to the gauge manufacturer. (Sizes can be obtained from the appropriate British Standards).

3. Design a suitable gauge for inspecting the centre distances between the 30 mm and 25 mm diameter holes shown in Fig. 189, Chapter 13.

4. Show, by means of sketches, a suitable receiver gauge to check the 100 mm height and lateral displacement from the Vee-location of the tailstock body shown in Fig. 192, Chapter 13.

5. A large number of splined shafts have to be given a 100% inspection. Sketch and describe the types of gauges that should be used to ensure that all the dimensions of the shafts are within the desired limits. Assume that the shaft has six splines.

(C.G.L.I.)

CHAPTER 11

Press Tools

The production of large quantities of cheap, sheet metal products, by means of press tools, is important in every sphere of life. The size and variety of the products of this manufacturing method range from common articles such as safety pins, cooking pans and car bodies, to driving chains, lamp housings and numerous component parts in office, domestic and other machinery.

The presses in use are hand- or foot-operated machines and fly presses for light work, e.g. umbrella frames; medium mechanical and hydraulic machines of various capacities for the majority of work, and heavy duty hydraulic presses in which large components such as car bodies are pressed. Other varieties of press, such as those used for hot pressing operations, are outside the scope of this chapter.

All presses have two parts of importance. One is the *platen*, which is stationary and carries the fixed portion of the press tools, known as the *die*; the other is the *ram* which carries the moving part of the press tool, known as the *punch*. The ram is caused to move in its guides by mechanical or hydraulic means.

The operations which can be carried out on press tools include blanking, piercing, bending, forming, drawing and coining.

Blanking and Piercing

Blanking is the process of punching out the sheet metal to any shape, the metal which remains being the scrap.
Piercing is the process whereby the formation of the hole is the desired result, the piece punched out being the scrap.

Tools should always be designed with regard to the numbers to be produced. Fig. 75 shows a set of tools suitable for producing approximately 12 500 blanks. The punch, which carries the shape being punched, is fitted into the ram of the press by means of a punch holder. The bolster, which carries the die, is clamped on the

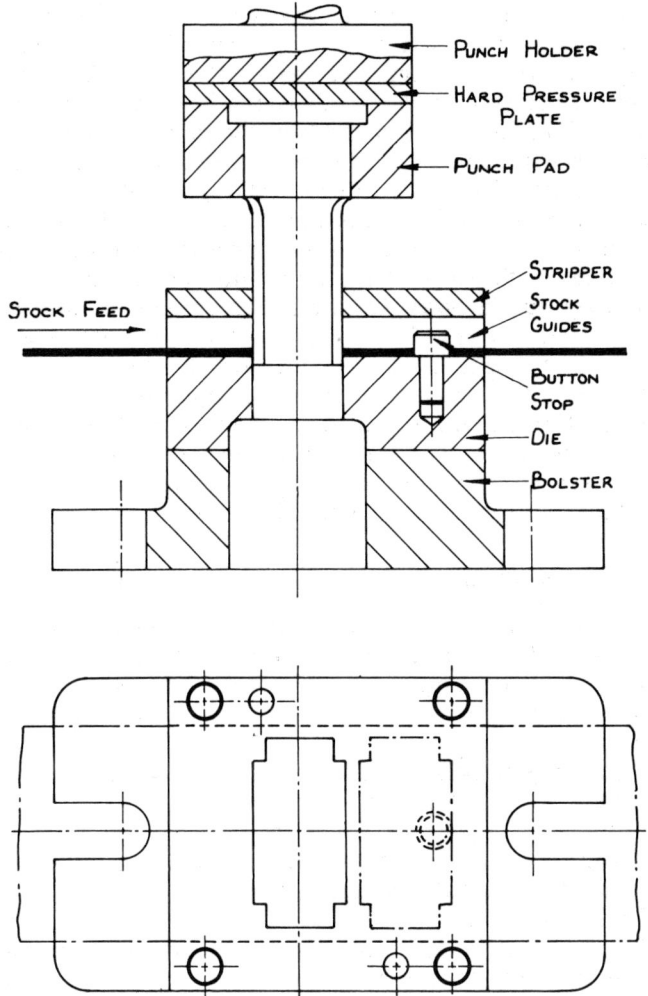

Fig. 75 Blanking tool for producing approx. 12 500 blanks

platen of the machine. The hole in the die is slightly larger than the punch. Stock guides are screwed and dowelled to the top surface of the die so that the distance between them is slightly wider than the stock strip being used for the punching process. Their purpose is to keep the stock strip in the correct position relative to the punch and die. The stripper plate is mounted on top of the stock guides and the thickness of the latter is such that gap between the die face and the underside of the stripper is sufficient to allow free passage of the stock strip over the button stop. The stripper removes the stock from the punch after a hole has been punched. When the hole has been punched from the stock material, the hole becomes slightly smaller whilst the piece punched out becomes slightly larger. This is due to the released compression of the crystals of the metal. The stock thus tends to grip the punch and the stripper prevents the stock from rising with the punch. After being stripped, the stock is moved forward so that a new piece can be punched out. In order that the correct amount of stock be fed into the die, a simple button stop is fitted into the assembly. During the stripping part of the operation, the stock is lifted clear of the stop. At the instant the stock falls away from the punch it is pushed forward. It also falls back towards the face of the die. The stock is thus progressed until the front edge of the hole has passed over the stop. The stock is then pulled back against the stop. Each successive hole is 'hooked' over the stop.

Punch and Die Clearances

A clearance is always provided between die and punch. This factor has considerable influence on the life of the tools and the power consumed during the punching operation. All designers have their own system of determining clearance, many are rule of thumb methods.

Suggested clearances for the common materials are:

Steel: 1/20th metal thickness;
Brass: 1/40th metal thickness;
Copper: 1/50th metal thickness;
Aluminium: 1/60th metal thickness.

The clearance is applied 'all round', i.e. when a circular blank is being cut, the difference in diameter between punch and die is equal to twice the calculated clearance.

The question arises as to whether the clearance is put on the punch or the die. When the blank being cut is the product required (blanking), the die is made the correct size and clearance is allowed on the punch.

When the blank being cut is the scrap (piercing), the punch is made the correct size and shape, and clearance is allowed on the die, i.e. the *punch* controls the *hole* size and the *die* controls the *blank* size.

Correct angular clearance is also put on the die; this will also lengthen the life of the die. Not more than two blanks should be capable of being held in the die. Fig. 76 shows suitable die clearance angles. No angular clearance is put on the punch unless a stripper block is not being used as this facilitates removal of the stock from the punch.

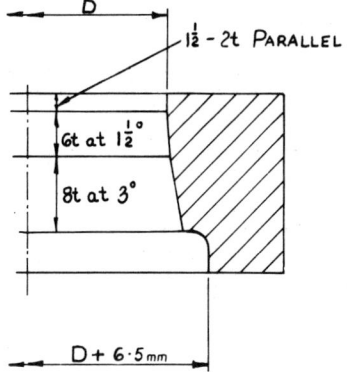

Fig. 76 Die clearances

Shear

The force required to punch a blank from the stock strip is dependent upon the ultimate shear stress of the material being worked, the thickness of this material and the length of the perimeter of the blank, such that:

blanking force = ultimate shear stress × thickness × perimeter of blank.

If, on determining this force, it is found to be nearing, or in excess of the maximum force that the press can deliver, the required blanking force can be effectively reduced by up to 25% by putting shear on the punch or die. Fig. 77 shows the method of putting shear on dies and punches. This shaping of the cutting edges of the tools reduces the amount of perimeter being cut at any one time, which in turn reduces the blanking force required. If a flat blank is required, shear is put on the die, but if the hole is the required product, the shear is put on the punch. When shear is put on the die, care must be taken in its design, otherwise location of the stock may be difficult, or the shear may interfere with easy movement of the stock over the die.

Punch and Die Holding

Punches are usually held in a punch holder which fits into the ram

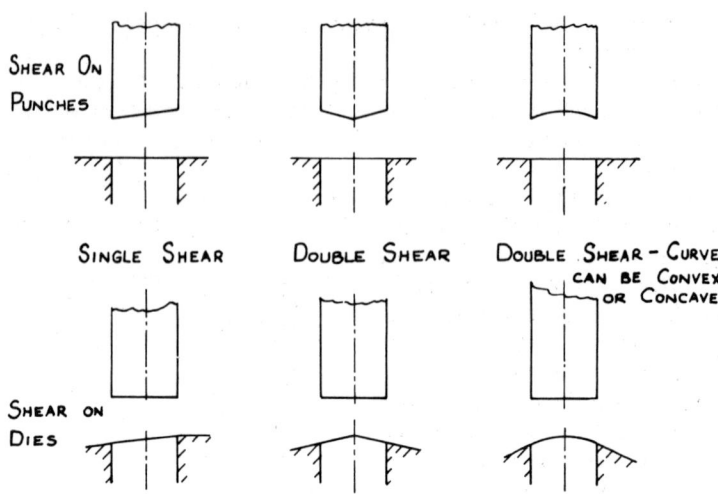

Fig. 77 Shear on punches and dies

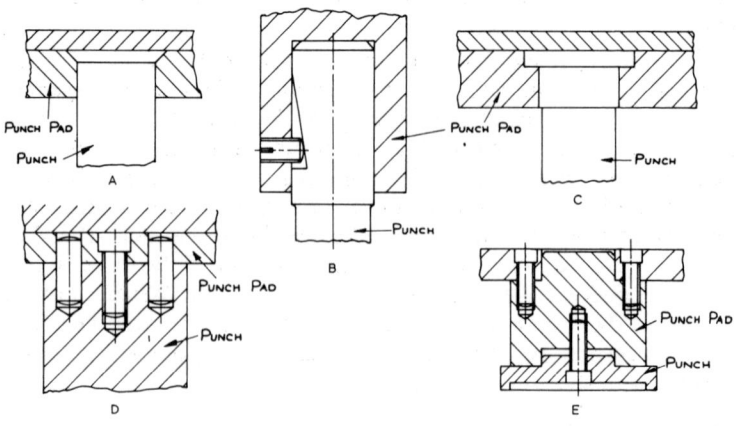

Fig. 78 Methods of holding punches

of the press. Fig. 78 shows some common methods of holding punches. In all cases the punch is assembled into the punch holder by means of a punch pad and is backed by a hard pressure plate (see Fig. 75).

Fig. 78A shows a punch which has been riveted over in the punch pad. A locking screw which clamps on to a flat, machined on the punch is shown in Fig. 78B, but this has the disadvantage that the screw may work loose. In Fig. 78C, a collar is machined integral with the punch. When large punches are being used holding systems shown at Fig. 78D and E are used.

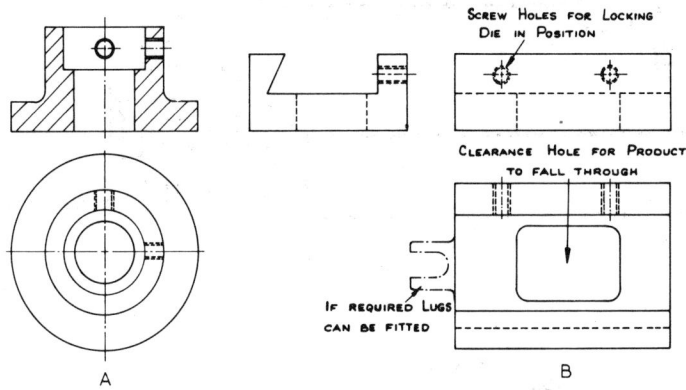

Fig. 79 Bolsters

The die assemblies are normally held in bolsters which in turn are clamped to the press platen. Fig. 79A shows a bolster used for replaceable circular dies whilst Fig. 79B shows a bolster suitable for flat dies.

Fig. 80 shows a die-set, many varieties of which are commercially available. It comprises a punch holder and bolster which are aligned to run together by means of guide pins or pillars. The punches and dies are mounted to the top and bottom members of the die-set and once correctly positioned, need not be disturbed, even when the assembly is removed from the press. Remounting of the tools back in the press is also simplified.

Stops

In order that the correct amount of strip is fed into the tools some device is necessary so that sufficient material to make a blank is present when the punch descends. On the other hand, the amount

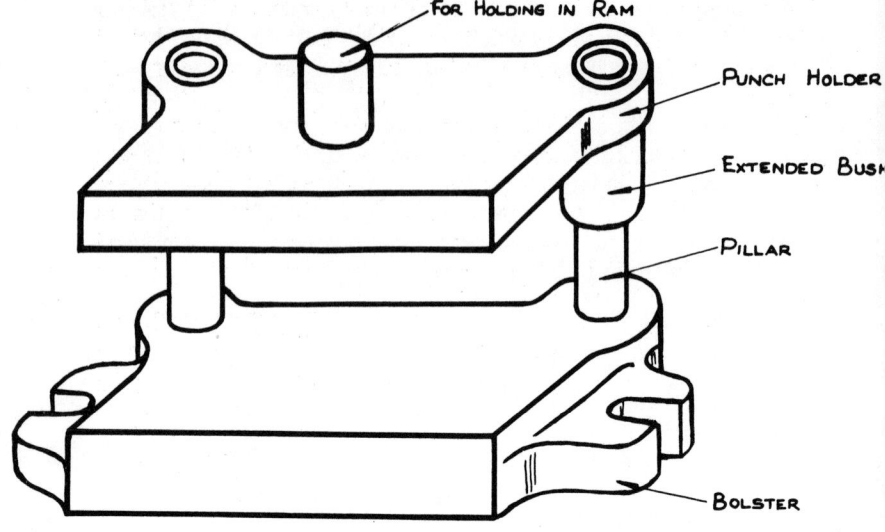

FOR HOLDING IN RAM

PUNCH HOLDER

EXTENDED BUSH

PILLAR

BOLSTER

Fig. 80 Die set

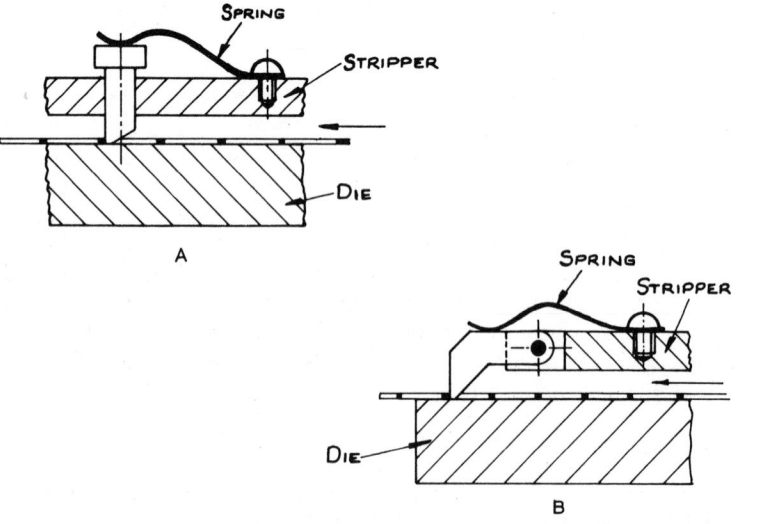

SPRING

STRIPPER

DIE

A

SPRING

STRIPPER

DIE

B

Fig. 81 Strip stops

of scrap stock should be kept to a minimum. These requirements insist that the same amount of stock should be fed into the tools after each stroke of the press ram. The simple button stop shown in the assembly in Fig. 75 is used when the stock is being fed by hand. Fig. 81 shows two semi-automatic stops also suitable for hand feeding operations. At (A), a spring loaded toe stop is shown whilst at (B) a pawl stop is illustrated. In both cases, the stock is pushed forward against the stop which rises under the pressure of the spring until the stop falls back into the hole which locates against the stop, when it is ready for the next stroke of the press. Operators can quickly acquire a good working speed.

Many different forms of stop have been designed, some allowing faster operation of the press than others, but the ultimate is continuous action of the press with the correct amount of stock being fed into the tools automatically. Many designs of automatic feeding devices are available commercially. These consists of mangle rolls which grip the strip and feed it into the tools. The rolls are driven from the crankshaft of the press; an adjustable eccentric pin on the crankshaft is connected to a rachet and pawl on the feed rolls.

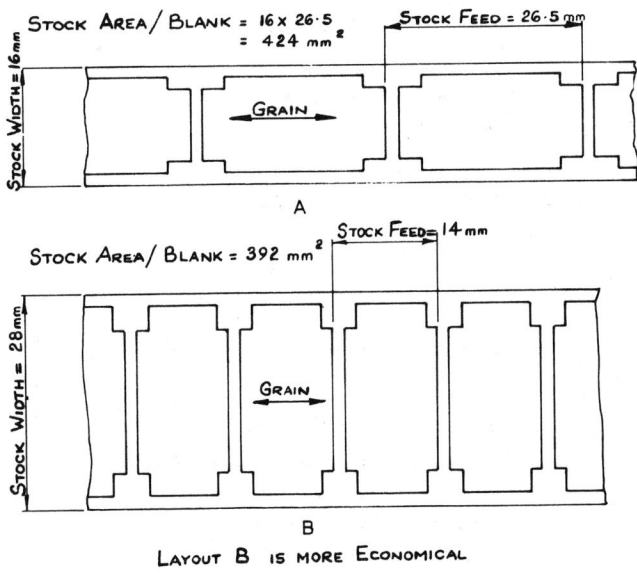

Fig. 82 Stock layout

117

The adjustment of strip feed is made by varying the radius of the pin on the crankshaft. When automatic feeding mechanisms are being used, stops are unnecessary.

Multiple Punches

It is often convenient for more than one blank to be punched out at each stroke of the press. Any reasonable number of punches can be mounted in the punch holder, the only consideration being the total force required for the shearing action. The layout when one punch is used is shown in Fig. 82, but the width of the stock will depend on which way the 'grain' of the stock is to run through the component.

When multiple punches are being used, economy in the use of material can be affected by careful layout. The aim is to produce as little scrap as possible. Fig. 83 A and B, shows economical

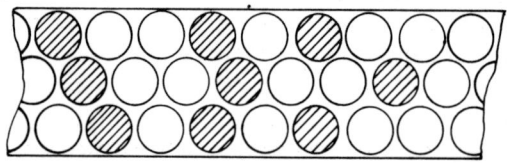

A   STRIP LAYOUT FOR CIRCULAR BLANKS
SHOWING 3 ALTERNATIVE PUNCH
ARRANGEMENTS.

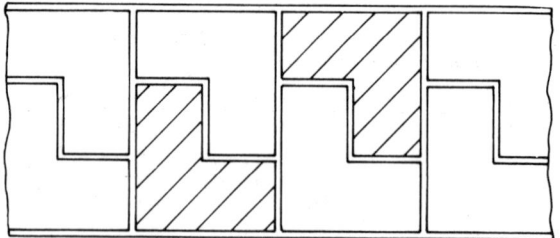

B   THIS LAYOUT COULD BLANKED BY USING TWO
PUNCHES OR BY USING ONE PUNCH AND
RUNNING THE STOCK THROUGH THE
DIES TWICE.

Fig. 83 Stock layout for multiple punching

layouts for multiple punching, the shaded portions being the position of the punches.

Follow-On Tools

A common method of producing a component which has got a hole in it, is by the use of a set of follow-on tools. Fig. 84 shows a set of tools for the production of a flat link. As the strip is fed through, the holes are first pierced and then the link is blanked out as the following holes are pierced. The stock layout is also shown in Fig. 84.

Bending

Together with blanking, bending is one of the most common presswork jobs, and like blanking, the cost and complexity of the tools should be directly related to the quantities to be produced.

Flat sheet and wire are bent into various forms by means of a male punch forcing the sheet or wire into a female cavity (die).

The radius of the bend should be as large as possible as this prevents cracking of the material and the bend should be made whenever possible, across the grain of the material, i.e. at right angles to the direction in which it has been rolled.

Bend allowances

Before a bent product can be made, the correct development of its

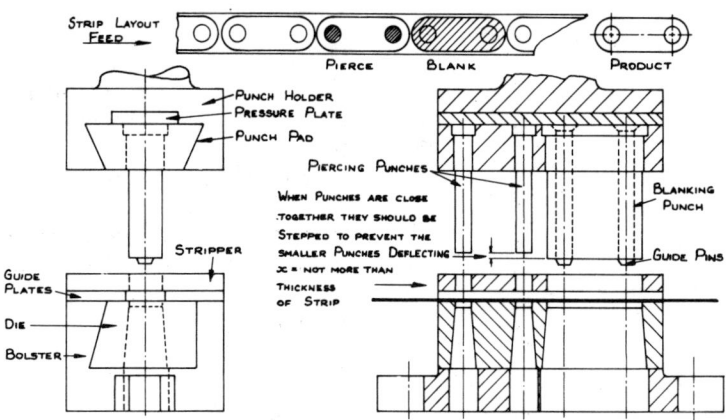

Fig. 84 Follow-on tools

BEND ALLOWANCES FOR MILD STEEL — 90° BENDS.

MAT⁺ THICKNESS		RADIUS OF CORNER (MILLIMETRES).												
S.W.G.	PREFERRED METRIC.	1	2	3	4	5	6	7	8	9	10	11	12	13
10	3·15						6·40	6·83	7·26	7·69	8·12	8·55	8·98	9·4
12	2·50					5·18	5·61	6·04	6·47	6·90	7·33	7·76	8·19	8·6
14	2·00				4·14	4·57	5·00	5·43	5·86	6·29	6·72	7·15	7·58	8·0
16	1·60			3·23	3·66	4·09	4·51	4·94	5·37	5·80	6·23	6·66	7·09	7·52
18	1·25		2·38	2·80	3·23	3·66	4·09	4·52	4·95	5·38	5·81	6·24	6·67	7·10
20	1·00	1·64	2·07	2·50	2·93	3·36	3·79	4·22	4·65	5·08	5·51	5·94	6·37	6·8
22	0·80	1·40	1·83	2·26	2·69	3·11	3·54	3·97	4·40	4·83	5·26	5·69	6·12	6·5
24	0·63	1·19	1·62	2·05	2·48	2·91	3·24	3·67	4·10	4·53	4·96	5·39	5·82	6·2

Fig. 85 Table of bend allowances

blank should be determined. The length of the blank will include the lengths of the straight portions together with the length of metal used in the bends. The use of the table of bend allowances (Fig. 85) is recommended for the speed and accuracy of determining blank lengths. When using the table, the total length of the straight portions of the bent product is determined; from this are subtracted the figures for the bends corresponding to the metal thickness and radius of the bend.

Example:

Fig. 86 shows a small bent product whose blank length is required.

Length of straight portions = $18 + 12 = 30$ mm

From table, bend allowance for radius and metal thickness = 4.09 mm

Blank length
$= 30 - 4.09 = 25.91$ mm

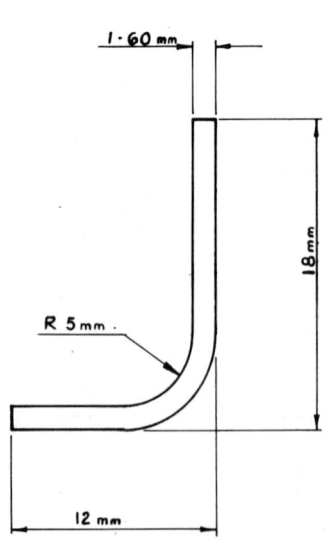

Fig. 86 Small bent product

Bending Tools

One of the simplest forms of

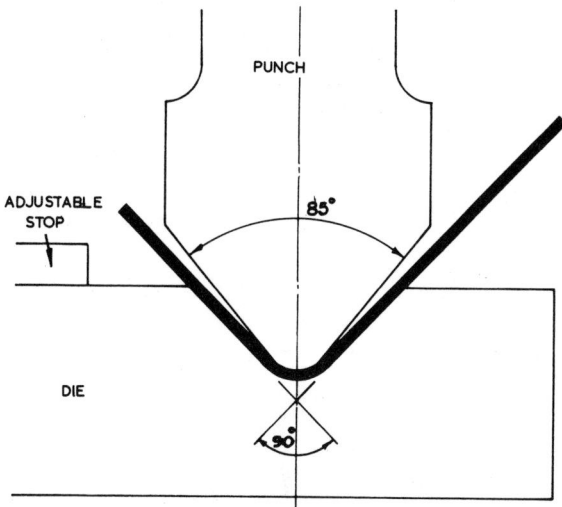

Fig. 87 Vee type bending tool

bending tool is the vee type. This can be made to bend any angle from 20° to 180°. When the stock has been positioned, the only other consideration is 'spring-back'. A satisfactory method of overcoming this difficulty is to make the punch with a sharper angle, e.g. if a 90° bend is required, the nose angle of the punch is 85° (see Fig. 87). This method allows the punch to give a solid blow to the product and consolidate the grain.

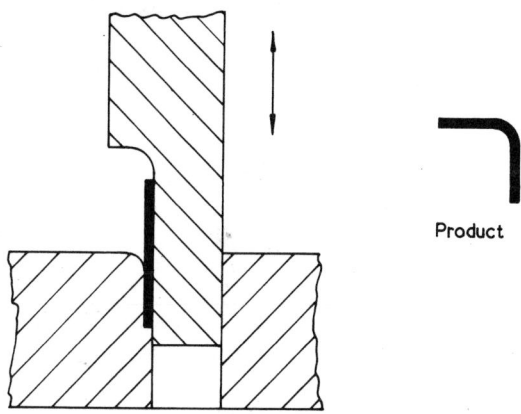

Product

Fig. 88 Simple bending tool

121

Fig. 88 shows another set of simple tools for making a bend in a small component. If two bends are required to be put into a component, a vee tool can only be used if the bends are far enough apart not to foul the tools, otherwise a box tool should be employed.

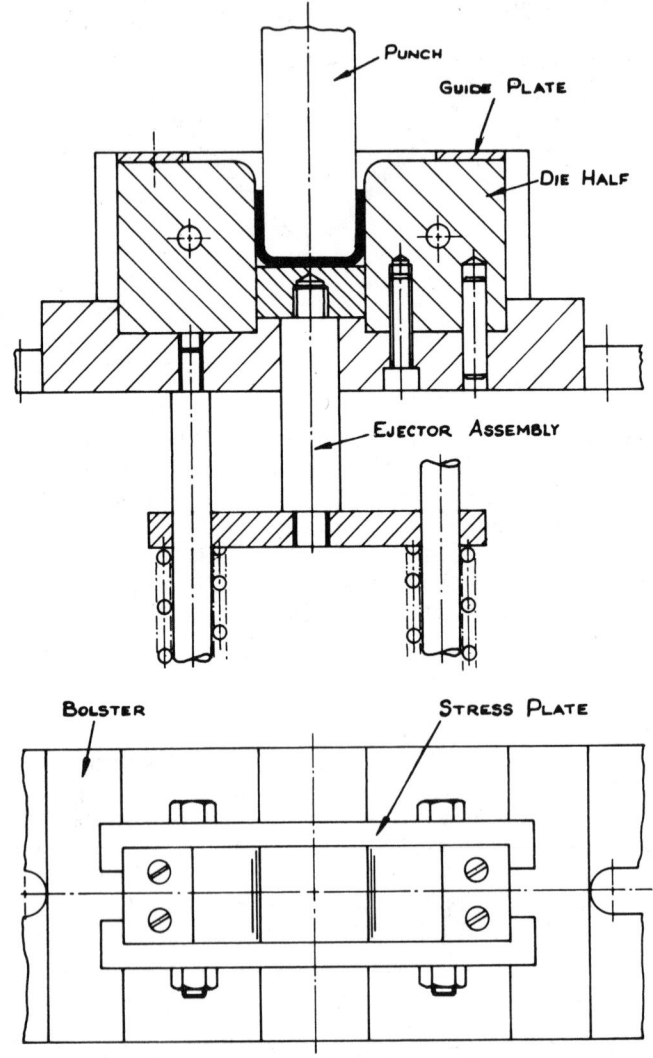

Fig. 89 Box type bending tools

Fig. 89 shows a typical set of tools for this work. On this type of tool an *ejector* is required to push the component from the die.

When a component is more complex, two or more operations may be required to produce it. Fig. 90 shows how a product is formed in two stages.

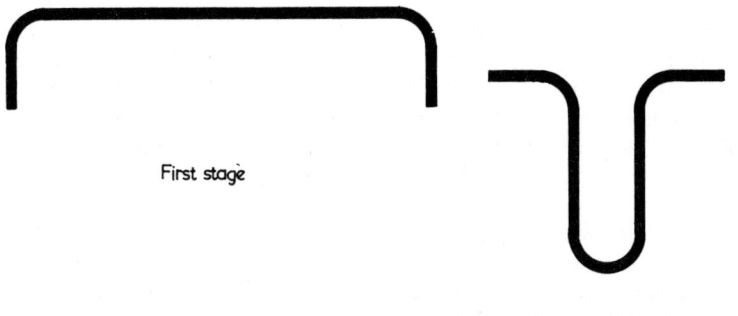

First stage

Second stage   finished product

Fig. 90 Product formed in two stages

Clearances on Bending Tools

In order that the tools may operate satisfactorily, correct clearances should be put on them.

Distance between sides of a bending die = overall width of component + 0·05 mm.

Distance over sides of a bending punch = overall width of component − 2 × metal thickness − 0·05 mm.

Press Tool Materials

A suitable steel for blanking and piercing punches and dies will depend on the quantities to be produced. For small quantities, a plain 1% carbon steel which has been hardened and tempered at 200°C, can be used but a generous grinding allowance should be made. When a long run is envisaged, the tools should be made from a steel containing 1·5–2·0% carbon and 12–14% chromium together with small amounts of other constituents. A suitable heat-treatment for this material is to harden at 960°C and oil quench, and temper at 220°C.

Materials for bending punches and dies also depend on the quantity to be produced. When small quantities are being produced, mild steel punches with zinc-base alloy dies can be used. When larger quantities are required, case-hardened mild steel tools will give excellent service. For long runs, tools should be made from a steel containing 1·5–2·0% carbon and 12–14% chromium.

Safety

Press tools require more efficient guarding than any other production process. Even with automatic feeding, some guard arrangement is required whilst the stock is initially fed in. Moreover, guarding is essential when hand feeding strip stock or component blanks. All guard arrangements should ensure that the operator's hands are clear of the danger zone during the operation of the press.

Guards can be classified as follows:

1. Fixed guards, which are permanently fastened to the press and cannot be moved by the operator; these never allow the operator's hand into the area of danger.

2. Interlock guards, which lock the clutch mechanism whilst the operator's hands are in the danger area. The press cannot be set in motion until the operator has closed the guard, and the guard cannot be closed until his hands are clear of danger.

3. Automatic guards, which 'remove' or 'sweep out' the operator's hands from the danger area during the movement of the press ram, are subject to severe regulations and H.M. Factory Inspectors ensure that the regulations are enforced.

EXERCISES

1. The clamp shown in Fig. 91 is to be made on press tools in two stages. The first stage is to produce the flat blank, and the second to make the bends.
   (*a*) Make a dimensioned sketch showing the approximate shape and sizes for the blank before bending.
   (*b*) Explain, with the aid of sketches, the general arrangement and operation of a press tool suitable for producing the blank.

(C.G.L.I.)

2. Using suitable sketches and explanation where required, describe the purpose and action of
   (*a*) the stripper plate of a blanking tool;
   (*b*) any form of self operating press guard.

(C.G.L.I.)

3. The component shown in Fig. 92 is to be made in quantity. Design suitable blanking tools for its manufacture. Establish a stock layout for maximum economy of material. The arrangement drawing of the press tools should carry a materials and parts list.

4. Fig. 93 shows a small bent component. Design press tools for:
   (*a*) blanking and piercing the flat blank;
   (*b*) bending the blanks in pairs.
   A materials and parts list should accompany each drawing.

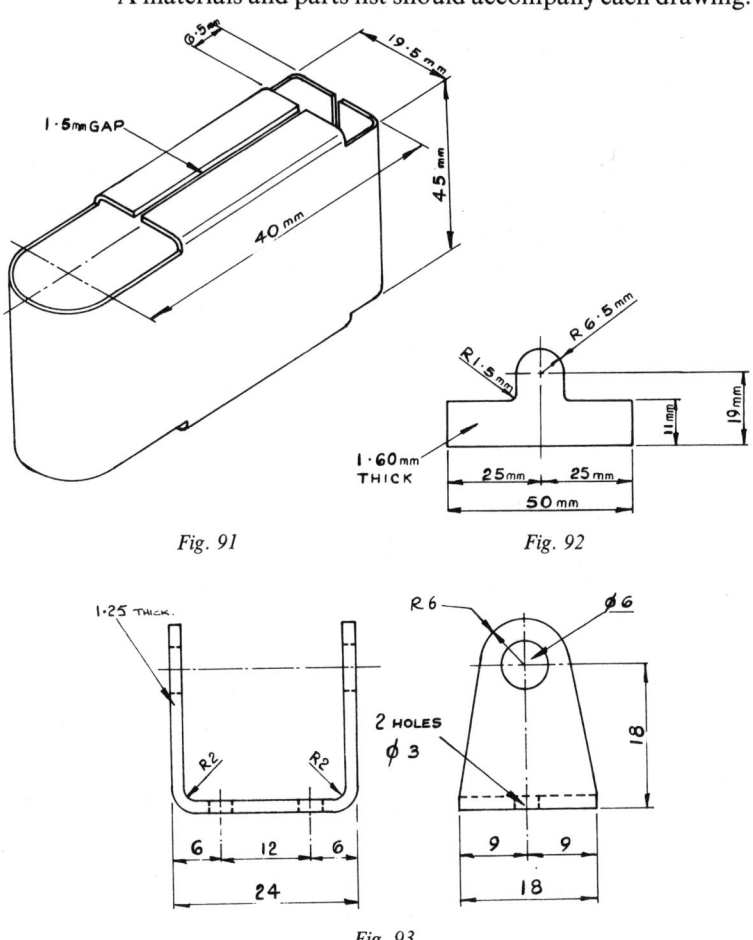

Fig. 91

Fig. 92

Fig. 93

CHAPTER 12

Jig and Fixture Design

Quantity production of engineering components which are accurate, interchangeable and relatively cheap is the desire of every production engineer. Where machine tools are concerned, the use of specially constructed jigs and fixtures will speed up production by reducing the amount of setting-up time required before the machining of the component can begin. A machine tool is only useful to a manufacturer when it is actually cutting metal; when it is idle, as during setting up, the cost of each component rises. Jigs and fixtures are used to reduce this idle time to a minimum by reducing the time between cutting one component and cutting the next. Indeed, it is often possible to arrange matters so that while one component is being machined, another is being loaded into the jig or fixture on the machine. The jig or fixture, because it holds all the work pieces identically and presents them to the cutting tools in exactly the same way, ensures that all components produced from that jig or fixture are exactly alike. This means that accurate, interchangeable parts can be produced. It is highly desirable that parts in assemblies, which are liable to wear, can be replaced when necessary with new parts manufactured to the same accuracy as the original.

Naturally, it would be folly to design and manufacture special jigs and fixtures if only a few components were required: the cost of tooling would be greatly in excess of the profit made on the components. On the other hand, if large quantities of components are required, it would be equally unwise not to provide jigs and fixtures. The cost of special equipment should be proportional to the quantities of components to be produced.

Jigs and fixtures are devices for holding a particular component for a specific machining operation. They are fitted with all the necessary equipment for clamping the component and for ensuring that each component is located in exactly the same position in or on the jig or fixture. Jigs are fitted with guides for the cutting tools. These guides are invariably hardened steel bushes through which the cutting tools pass and can rotate. Jigs are used mainly for drilling and boring operations. A fixture does not guide the tools, but simply holds the work in the correct relative position on the machine tool. The work and cutters are brought into operational contact by means of the normal machine movements. Fixtures are generally used for milling, grinding, turning and broaching.

General Considerations

Each component has its own distinctive features such as shape, weight and size, and will require a jig or fixture of a different design to enable it to be machined. Whilst the design problems vary with each particular component there are several factors which are common to all jigs and fixtures. These factors require consideration at the design stage and are outlined below:

1. Rigidity. Jigs and fixtures should be sufficiently rigid to withstand the vibrations and forces present during cutting. Milling fixtures in particular, should be massive enough to absorb vibrations otherwise chatter between work and cutter will result.

2. Holding the Fixture to the Table. When the fixture is fastened to the machine table it should be bolted solid and not sprung out of alignment in any way. To facilitate alignment of the fixture with the cutters, tenon keys should be fitted to the machined underside of the fixture. These will fit into the tee-slots of the machine table and locate the fixture. A centre plug may be of use if the fixture is to be used on a rotary table or lathe faceplate.

Generally speaking, jigs are not bolted on to the machine table but this is not always so: all jigs used on boring machines must be bolted to the machine but a drilling jig for a component in which there are a number of similar holes, would not be bolted down when used on a single spindle drilling machine. However, if a multi-spindle drill head is to be used, the jig would be bolted down.

3. Foolproofing. Often, components are so shaped that it is impossible to load them into a jig or fixture in the wrong position, so that a faulty cut may be taken. The jig or fixture should be designed in such a way that it is impossible to load the component in any but the correct position. A pin or abutment can usually be strategically placed so that it fouls a component which is being wrongly loaded.

4. Loading the Component. It is essential that the component can easily be fitted into the jig or fixture. If the component is heavy, it should be permitted to slide into the jig without difficulty.

5. Location of the Component. All location points should be clearly defined and designed so that they are not swarf traps. If a hole has been chosen as a location, it should be ensured that it has a close tolerance. If the location must be from a hole having a wide tolerance, arrange for the tolerance on this hole to be tightened, so that it will ensure accuracy on subsequent operations.

6. Clamping. All clamping and adjusting operations should be carried out on the side of the fixture nearest the operator and the fixture should be designed with this factor in mind. On milling fixtures, fixed stops should be arranged so that they, and not the clamps, take the direct thrust of the cut. If the component has some finished surfaces and it is essential that these surfaces do not become marked by the clamps, the latter should be fitted with soft inserts of fibre. Clamping parts should be arranged so that they act directly above the work support points, otherwise the component may be sprung and disturbed. Clamps and other operational features of the jig or fixture should be constructed so as to avoid the use of spanners wherever possible. When this cannot be done all nuts should be the same size, obviating the necessity for spanners of different sizes.

7. Coolant and Swarf. A plentiful supply of coolant should be available at the cutting point to wash away all the swarf into the base of the jig or fixture and then out on to the machine table. Holes should be provided in the main body of the jig or fixture to allow free passage of coolant and swarf. Jigs and fixtures should be carefully designed so that they do not become swarf traps.

8. Ejection of the Component. When machining is complete, the component must be removed from the jig or fixture and it is essential that this can be done with ease. Burrs at the edges of a machined surface may prevent the component from being withdrawn easily and it may be necessary to provide burr grooves. On heavy components, some type of mechanical device may be useful for the ejection of the component.

9. Safety. At all stages of design, the safety of the operator should be considered; every precaution should be taken to ensure safe operation of the jig or fixture.

Jig and fixture details

The general details of jigs and fixtures have to some extent become standardized over many years of experience. Indeed, many firms now produce standard parts for jigs and fixtures. These enable a designer to construct a jig or fixture which requires a minimum of machining, the bulk of the design being made up from standard parts; the time for design and manufacture of the jigs or fixtures is reduced and the equipment is ready for the production line much quicker than when each part is made individually. The standard parts can be kept in stock so that the replacement and repair of jigs and fixture is made quickly and easily.

Figs 94–110 show what has become good practice in the way of jig and fixture details.

Drill Bushes

These are used in drilling and boring jigs for guiding the drills, reamers, counter bores and boring bars to the component. The cutting tool is thus located in the correct relative position and is guided for the whole of the cutting operation. Fig. 94 shows a plain drill bush; this is usually cheaper than any other type of

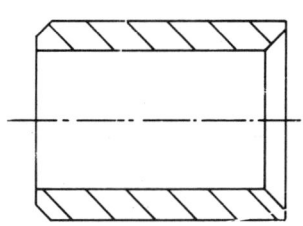

Fig. 94 Plain drill bush

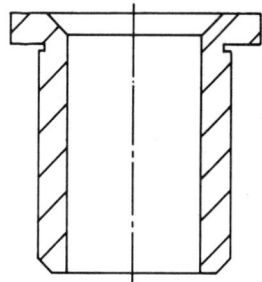

Fig. 95 Flanged drill bush

bush but may not be as satisfactory as the flanged drill bush shown in Fig. 95. The flanged type gives more bearing to the drill and is very useful when it is desired to feed down a drill to a dead stop. It is also more easily found by an operator when it is fitted into a jig plate likely to get covered with swarf. In a thin jig plate the flanged bush can still be used; by increasing the diameter of the flange to accept small countersunk head set screws, the bush can be held down. The flanged type of bush should be used in preference to the plain type but both should be a good press fit into the jig plate.

When a further operation is necessary on a drilled hole such as reaming or tapping, it is better to leave the component in the jig rather than remove it to load into another jig for the operation. Special removable bushes, called slip bushes, can be fitted to the first jig. The outside diameters of a range of slip bushes are identical, but the bores vary. After drilling a hole for reaming, the drill sized bush is removed and the reaming sized bush is fitted for the reaming operation. When a tapping operation is necessary, the slip bush is removed but not replaced, the tap being passed through the hole left by the slip bush. Because slip bushes are being constantly

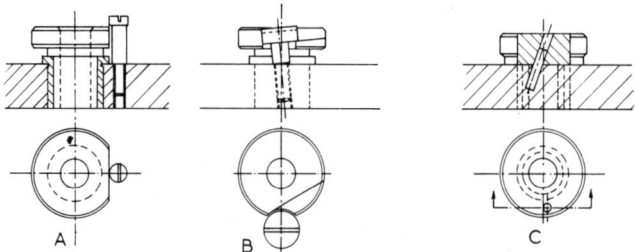

Fig. 96 Slip bushes A. Bush fitted in position with flat next to pin; given half a turn to prevent lifting. B. Similar but more costly method of locking. C. Simple method of holding slip bush; locking pin needs frequent replacing

removed and refitted to the jig, a liner bush is first fitted into the jig plate, the slip bushes fitting into this. Fig. 96 shows various methods of holding slip bushes in position. Fig. 96A shows the slip bush with a flatted flange; the flange is fitted to the jig with the flat next to the locking pin and the bush is given half a turn to prevent it from lifting from the jig.

Another similar method of locking the slip bush is shown in Fig. 96B. This bush is a little more costly but is more firmly held in the jig, by means of the taper locking system, when given half a turn. Fig. 96C shows a very simple method of holding a slip bush by means of an inclined pin which prevents rotation of the bush. The pin needs replacing at frequent intervals because of the extreme wear.

Special drill bushes can sometimes be used for location and clamping purposes. Fig. 97 shows a drill bush which does treble duty as a clamp, as a location and as a drill bush. The bush has a large tapered portion at the bottom, inside the bush, and this will

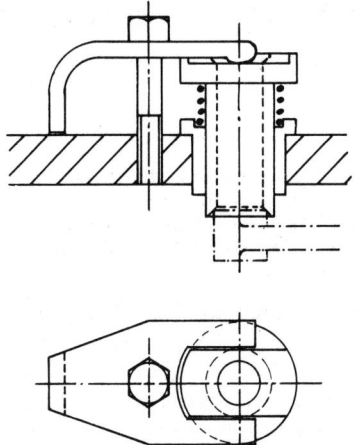

Fig. 97 Drill bush arranged for locating and clamping

locate a component having a circular boss such as the one shown in Fig. 97. As the bush cannot move, it will adequately locate the component at that point. The bush is also spring loaded and fitted with a clamp which allows the workpiece to be clamped in position. When clamping pressure is released, the spring raises the bush from the workpiece. Fig. 98 shows a bush acting as a clamp and as a drill bush. The bush is screwed down in its housing to clamp the component with the bush maintained in alignment in the housing by means of the fore and aft pilot bores. The screw thread must be a fairly loose fit for the bush to function correctly.

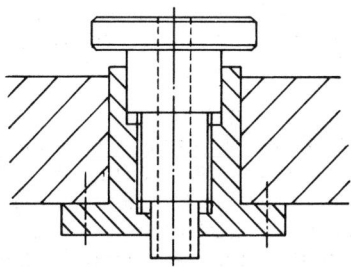

Fig. 98 Drill bush arranged for clamping

131

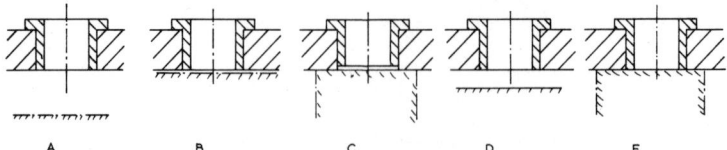

Fig. 99 Mounting of drill bushes A. Bush too far from work. B. Bush too close to work; gap likely to get crowded with swarf. C. Gap between bush and work will collect swarf and drill may break. D. Good practice. E. Good practice.
As a general rule the length of the drill bush should be equal to twice the drill diameter

Fig. 99 shows examples of good and bad practice in the mounting of drill bushes in jigs. Fig. 99A shows a drill bush mounted too far away from the workpiece. This may allow the drill point to wander over the workpiece before commencing to drill. The hole may not be in the correct position because of this. In addition, because the drill is not drilling square to the workpiece, the holes may not be square to the surface of workpiece; extreme wear of the drill bush and a broken or bent drill may also result. At Fig. 99B the bush is mounted too close to the work; the resulting space may become crowded with swarf and subsequently trap and break the drill. A similar result is likely to occur with bushes mounted similar to the one shown in Fig. 99C, with the additional hazard of the bush being forced out by the swarf. Fig. 99D shows good practice when mounting a drill bush having a space between it and the workpiece. The gap is neither too large nor too small and it fits flush to the jig plate. The gap may vary between $\frac{1}{3}$ to $\frac{1}{2}$ of the drill diameter but if a particularly heavy cut is to be taken, the gap can be increased. Fig. 99E shows good practice in mounting drill bushes flush to the surface of a component.

The length of drill bushes is also an important factor. Generally speaking the length of the bush should be equal to two drill diameters, but a shorter bush can be used on the larger diameter drills as these are much stiffer and less likely to deflect.

Jig Feet

Jigs which have to be moved about on machine tables must be provided with jig feet. Any swarf on a machine table would prevent the jig from standing firmly on the table if the jig had a flat base. The jig should stand on small feet set at the extremities of the jig so that the swarf is not so much of a problem. The feet, whilst they should be small, should not be so small as to fall through the slots

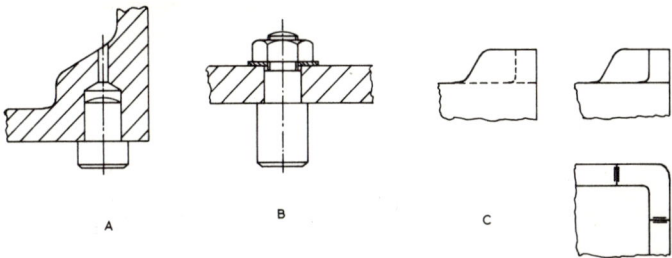

Fig. 100 Jig feet A. Foot pressed into cast jig base; provision should be made for removal of foot. B. Foot held in position by means of a nut and screwed portion on foot. C. Feet of this type are cast integral with the body of the jig

in the machine table. Fig. 100 shows three varieties of jig feet. At A a simple button foot is shown pressed into the base of a jig. If the hole is extended through the jig base, easy removal is facilitated, and air is allowed to escape as the foot is pressed home. The jig foot shown at B is held in place by a nut and washer; this enables easy removal and replacement. Both these types of jig feet are suitable for jigs of cast and fabricated construction but at C, the foot shown is suitable only for cast jig bodies as it is cast integral with the jig body itself. It has the disadvantage of not being able to be replaced when worn.

Locations

A body in space is free to move in any direction it chooses; it can rotate and can move laterally in any plane and so must be constrained when being machined in order to prevent any movement; this constraint is achieved by using a combination of locations and clamps.

Perhaps the most obvious location is a previously machined hole. Provided that the hole has been machined to a fairly close tolerance a locating pin can be used for location purposes. Fig. 101 shows three locating pin ideas in common use. At A, the pin is a press fit into the jig or fixture, the shoulder preventing it from being inserted too far. The pin is tapered to allow for easy loading of the component and the parallel portion of the pin is a sliding fit into the component. When a heavy component is being handled, the disappearing location, shown at B, can be used. The workpiece is first placed in the jig or fixture and the pin is then raised by means of the eccentric pin actuated by the hand lever. When two holes

133

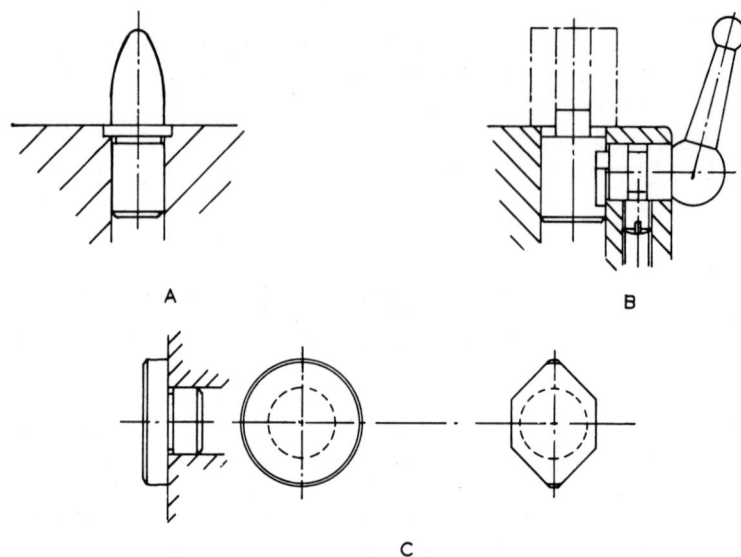

*Fig. 101 Locating pins. A. Tapered shouldered locating pin. B. Disappearing location.
C. Two pins for locating from two holes*

are being used for location purposes, two pins are used; the first
is a full pin, whilst the second pin has been flatted to give only two
small arcs, shown at C. This will allow a component to be located
when the centre distance is likely to vary by a small amount. When
hole locations are being used, the position of the hole must be
held to a close tolerance otherwise its use as a location cannot be
accepted. It often happens that one hole has to be used as a location
when, in fact, the tolerance on the hole is not satisfactory. In this
case the tool designer, should arrange for the tolerance on the hole
to be tightened to enable it to be used for location purposes.

When the component has a circular or partly circular external
form, a vee location can be used. Fig. 102 shows a fixed vee location
holding a circular bossed component. The gripping face is undercut,
so that swarf will not unduly affect the working of the location,
and angled back so that the location tends to hold the component
down on to the base of the jig or fixture. It is advantageous to have
the fixed vee location tenoned into the base of the jig to prevent
any knocks or shock loads disturbing it. If the free end of a component must be located in a vee, a sliding vee must be used. Fig. 103
shows a simple sliding vee location operated by a hand-operated

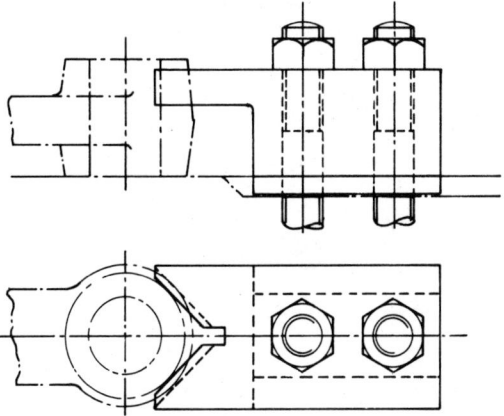

Fig. 102 Fixed vee location

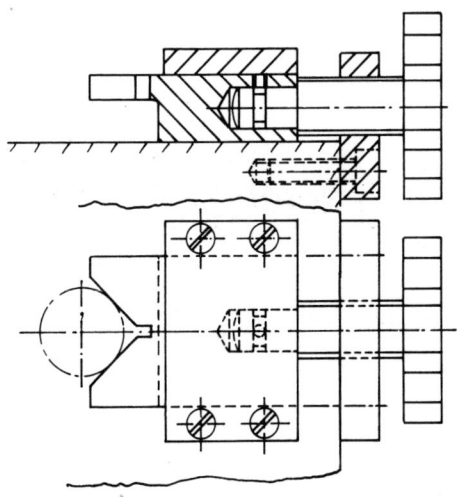

Fig. 103 Sliding vee location

screw. The screw is secured in the sliding vee piece enabling back and forward movement to be obtained. The vee piece is guided and held in the correct position by means of a coverplate. Fig. 104 shows other sliding vee arrangements operated by screw and cam. At A the thrust applied to the component is transmitted to the body of the fixture through the collar and screw thread; at B the

135

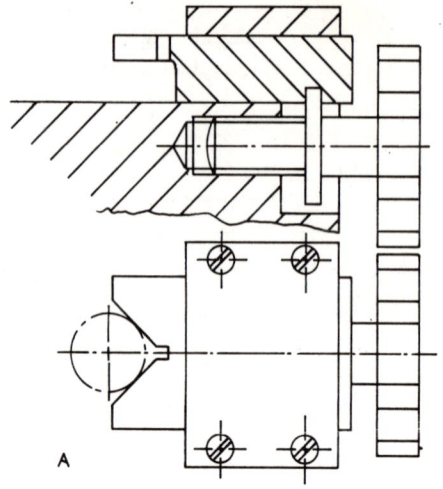

A

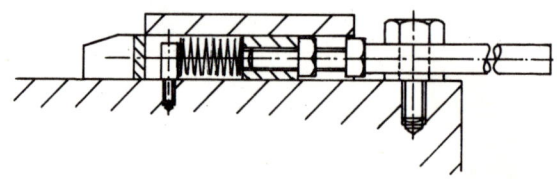

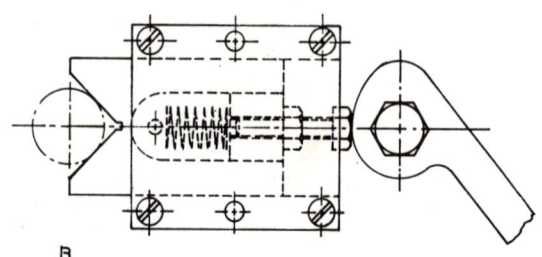

B

Fig. 104 Sliding vee locations

sliding vee location is more rapid in operation, being manipulated by a cam lever. The return of the vee is by means of a spring. The vees should always tend to hold down the component by having their gripping faces angled back.

Supporting Devices

A component should always be clamped at a point where the work is supported. As it is not always possible to clamp a workpiece at a point where it touches the walls and base of the jig and fixture, some supporting device is used for this purpose. When a component is likely to deflect under the cutting load, a supporting device can be used to overcome this.

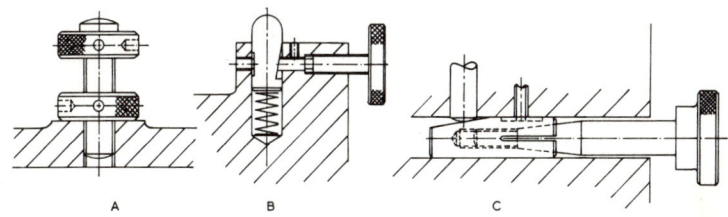

Fig. 105 Work supporting devices A. simple jack with locking collar. B. An auto-matic positioning pin. C. A wedge-actuated pin

Fig. 105A shows a very simple screw jack with a locking collar. The jack portion is raised by hand until it meets the workpiece when the locknut is screwed down. Tommy-bar holes are provided in each part. This method of support is rather slow but a faster operation is achieved by the use of the device shown at Fig. 105B. This is an automatic positioning pin. The weight of the component positions the pin against light spring pressure whilst the hand screw locks the pin in place after location; it is prevented from rotating by means of a small dowel. At Fig. 105C a wedge actuated pin is shown. The hand screw is pushed into the hole, the pin rises to meet the workpiece and the hand wheel is turned to lock the assembly in position. The locking action is obtained by the taper on the hand wheel shaft being forced into a mating taper in the wedge pin, which is prevented from rotating by means of a screwed peg. Fig. 106A shows a simple spring loaded support pin. Its use, how-ever, is limited, as only a small movement can be obtained. Fig. 106B and C shows locking devices for support pins. The better, but

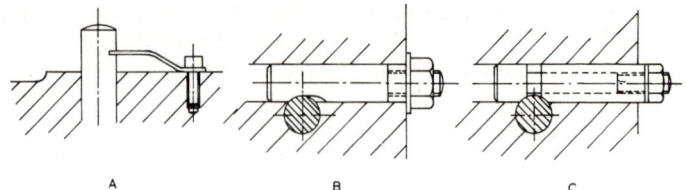

Fig. 106 Work supporting devices A. Support pin is spring loaded. B. and C. Locking device for jack pins

more costly design is that shown at C and it is similar to that used for clamping tailstock barrels in position. When the nut is tightened, the head of the bolt and the sleeve are pushed together, thus clamping the support pin.

Clamps

Most jigs and fixtures have incorporated in their design, some form of clamping arrangement, and its purpose is to hold the component in the position determined by the locations. All clamping should take place where the work is supported, otherwise springing and permanent deformation of workpiece will result.

Fig. 107A, B and C shows similar types of clamp, the differences between them being in the formation of the heel. At A, the clamp is supported on a separate heel pin; a groove and slotted hole in the clamp allows the clamp to be withdrawn from the work without rotating, and the light spring around the stud supports the clamp

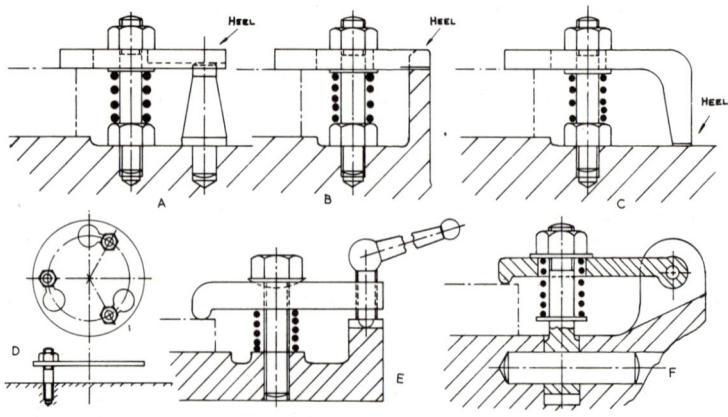

Fig. 107 Screw-operated clamps

to facilitate easy replacement on a new workpiece. At B, a similar arrangement is shown, but the heel is supported on part of the fixture body, suitably slotted to allow sliding of the clamp, but not rotation; the heel end of the clamp is radiused to give three point contact. At C, the radiused heel is made integral with the clamp and, as shown, no provision for preventing rotation of the clamp is made. All clamps on a jig or fixture requiring a spanner should have nuts of the same dimension across the flats.

Fig. 107D shows a clamp used regularly for clamping circular work. The stud holes are slotted and enlarged at the ends so that it can be slipped over the nuts without difficulty. This avoids having to remove the nuts.

When it is desired to clamp without having to use a spanner, the arrangement shown at Fig. 107E could be used but the effective clamping force is not as high as when spanners are used. The clamp pivots about the stud and a spherical washer (in a spherical seat) is required. To obtain three point contact, the end of the hand screw is made spherical and this runs in a slot to prevent the clamp from rotating. If the stud hole in the clamp is slotted, the clamp can be retracted from the work.

The clamp shown at Fig. 107F is used when both clamp and clamping screw must be moved clear when loading a component. The clamp and clamping screw are both hinged and the pivots for these must be accommodated in the jig and fixture. A light spring supports a washer under the nut to save the operator's time.

When rapid but not heavy clamping is required, the hand screw

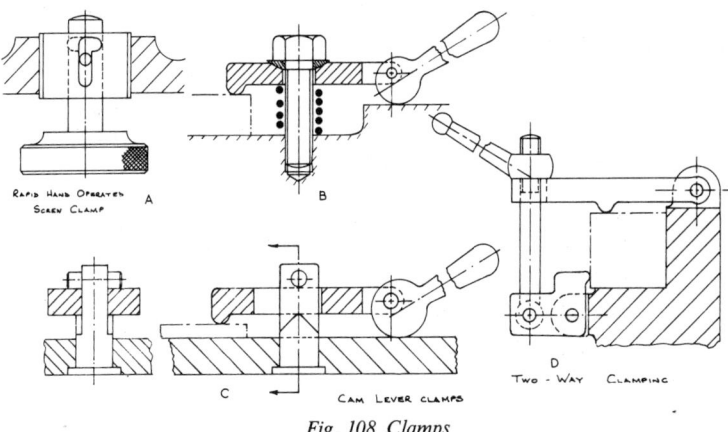

Fig. 108 Clamps

clamp shown at Fig. 108A can be used. A centre pin, fastened to the hand wheel, does the actual clamping. A smaller pin passing through the centre pin engages in slots cut into a screwed plug. In action, the hand wheel is pushed forward so that the small pin moves to the bottom of the slot, and at this point the hand wheel is turned in a clockwise direction so that the pin engages in the foot of the slot. Any further rotation of the hand wheel results in turning the screwed plug which moves the whole assembly forward to clamp the component. To release the component, anti-clockwise rotation of the hand wheel will turn the screwed plug back, when the small pin can be withdrawn to the top of the slot, thus allowing the clamp to clear the workpiece.

Cam operated clamps can be used to great advantage when rapid clamping is needed without the use of spanners, but again the clamping force is not so great. Fig. 108B shows a clamp similar to those described previously operated by a cam lever. Note the use of a spherical washer and seating at the stud. The clamp shown at 108C can be withdrawn from the work to facilitate removal of the workpiece. A pillar passes through a slot in the clamp and the cross pin keeps the clamp in place. The clamp is prevented from falling by the material left on the pillar during its manufacture.

When a rapid clamp is required to clamp in two directions at the same time an arrangement similar to Fig. 108D can be used, the clamps should apply equal pressures to the workpiece. The top clamp is slotted so that the clamps can be swung clear for manipulation of the workpiece.

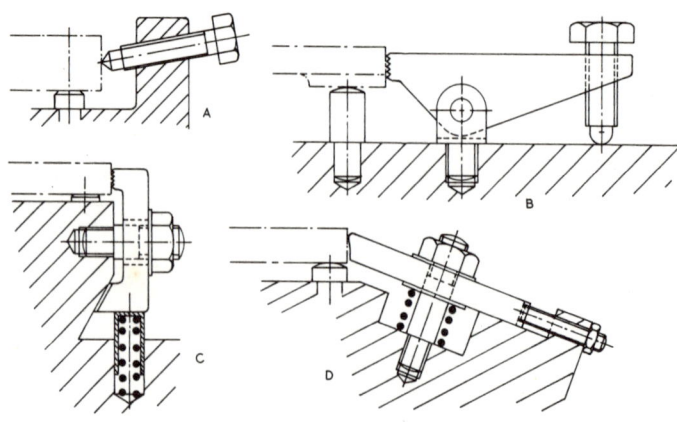

Fig. 109 Edge clamps

Fig. 109 shows several varieties of edge clamp. In all cases the work is clamped sideways and downwards at the same time. This type of clamp is very useful when no part of the clamp must protrude above the surface of the workpiece. At A a screw with a hardened point is used as a clamp. It is cheap, simple and very efficient.

The clamp shown at B works on the lever principle, but can only be used where there is little variation in length of the components. Larger variation in length can be tolerated by the clamp as shown at C, but a fairly large fixture depth is required. The spring loaded heel pin supports the clamp until it is brought into action by means of the nut. Because of the hardened points and serrations, A, B and C can only be used where the marks left by clamps are not an objection. The edge clamp shown at D does not have serrations and it is, to some extent, adjustable to accommodate varying lengths, but only if the variation is in batches.

In Fig. 110A another cam clamp is shown, the angled face of which holds the component down whilst the cam face pushes the component up against its locations.

The cam actuated clamp shown at Fig. 110B is useful when clamping space is at a premium. The clamp is shown situated beneath an overhanging component. When the screw is turned, the cam is pushed round to lift the heel of the clamp and pivot against the spherical seating of the pillar.

A captive clamping plate is shown assembled in Fig. 110C. It is often used on drilling jigs for components such as bushes and

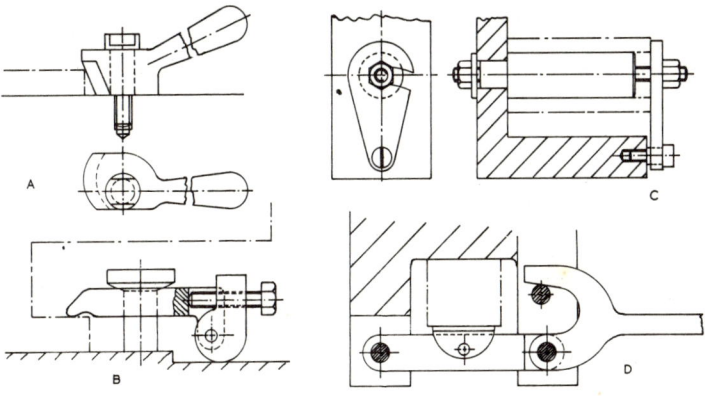

Fig. 110 Clamps

sleeves which have been previously machined but require such items as oil holes drilling in them. The component is pushed over a locating pin and held in place by the clamping plate which is swung up into position beneath the nut. Being fastened to the jig, the clamping plate cannot be lost. The use of a captive clamping plate is to be preferred to the use of loose slotted or C-washers which tend to be frequently lost.

The hook cam clamp shown in Fig. 110D, is ideal where only

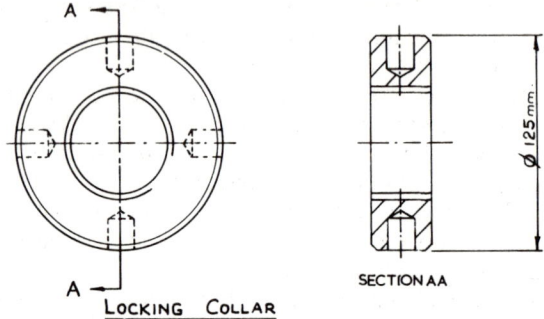

LOCKING COLLAR

SECTION AA

A JIG IS REQUIRED FOR DRILLING THE 4 Ø12 mm. TOMMY-BAR HOLES AROUND THE PERIPHERY OF THE COLLAR. THE COMPONENT IS RECEIVED FULLY MACHINED EXCEPT FOR THE 4 HOLES

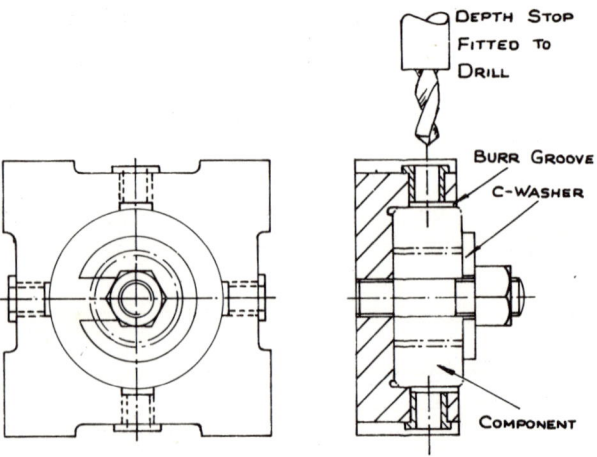

Fig. 111 Solid type jig for locking collar

light pressure is needed and where machining time is short, as it tends to shake loose during operation. In the figure, the clamp is fitted to a clamping bar carrying a floating pad which actually holds the work; pressure is applied by hooking the clamp over a pin fixed in the jig and fixture.

Design examples

Figs 111–115 show jigs and fixtures which have been designed for machining operations on various components. It will be seen that they comprise combinations and adaptations of standard details mounted on a suitable base unit.

Fig. 111 shows a drilling jig suitable for drilling the four tommy-bar holes around the periphery of the locking collar. The drilling of the holes is the final operation in the manufacture of this component and the collar is fully machined, except for the four holes, when it is received for this operation. The collar is located in the jig on its outside diameter and is clamped by means of a nut and C-washer; burr grooves are provided to accommodate any burrs thrown up during machining.

The component shown in Fig. 112 is a cast iron distance piece. The 50 mm diameter holes and the face 'A' have been machined and a fixture is required for gang milling the 25 mm wide tenon and associated faces, and a jig is needed for drilling the 8 × 13 mm diameter holes. Fig. 113 shows the general arrangement of the drilling jig with the component shown in chain-dot lines. The

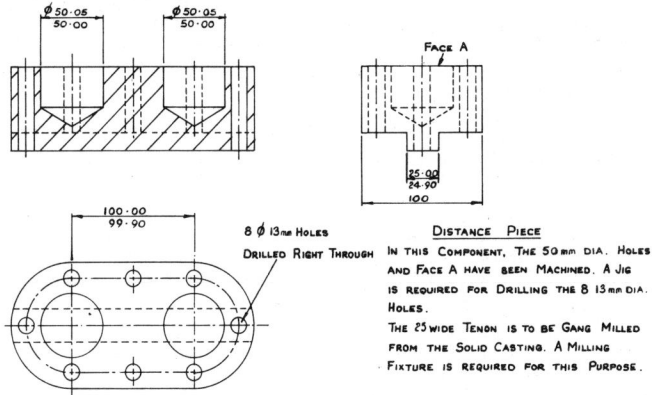

8 Ø 13mm HOLES DRILLED RIGHT THROUGH

DISTANCE PIECE

IN THIS COMPONENT, THE 50 mm DIA. HOLES AND FACE A HAVE BEEN MACHINED. A JIG IS REQUIRED FOR DRILLING THE 8 13 mm DIA. HOLES.

THE 25 WIDE TENON IS TO BE GANG MILLED FROM THE SOLID CASTING. A MILLING FIXTURE IS REQUIRED FOR THIS PURPOSE.

Fig. 112 Distance piece

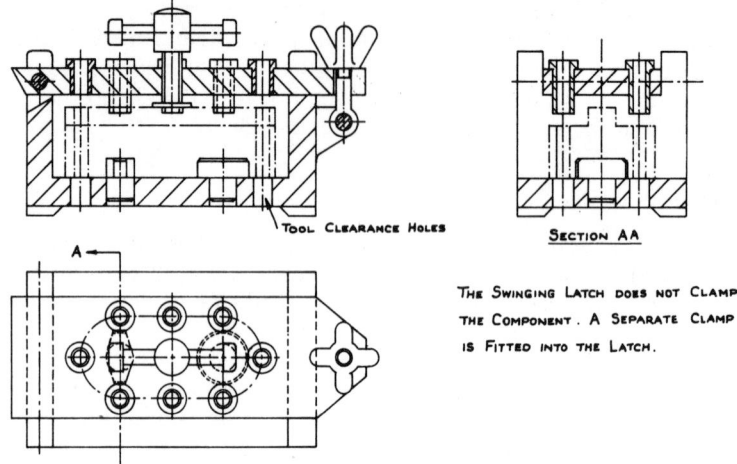

TOOL CLEARANCE HOLES

SECTION AA

THE SWINGING LATCH DOES NOT CLAMP
THE COMPONENT. A SEPARATE CLAMP
IS FITTED INTO THE LATCH.

Fig. 113 Box jig for distance piece

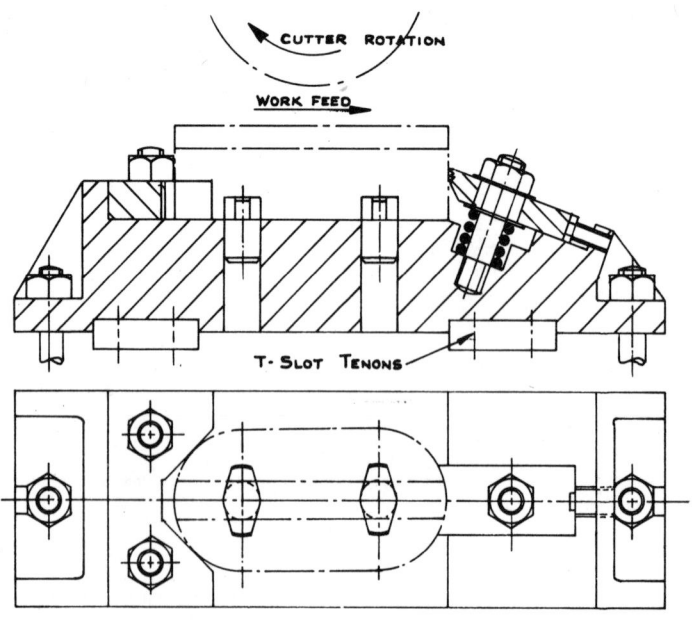

CUTTER ROTATION

WORK FEED

T-SLOT TENONS

Fig. 114 Milling fixture for distance piece

144

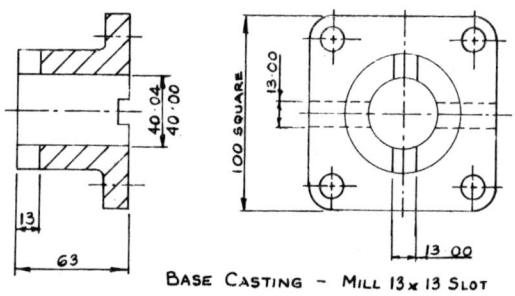

BASE CASTING ~ MILL 13 x 13 SLOT

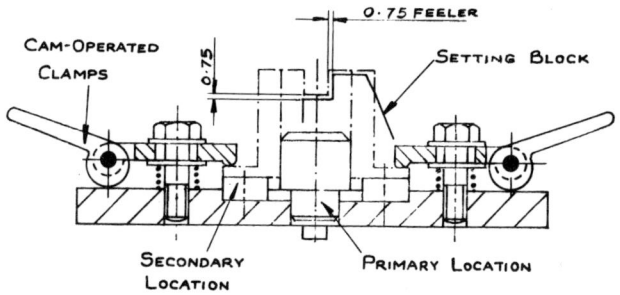

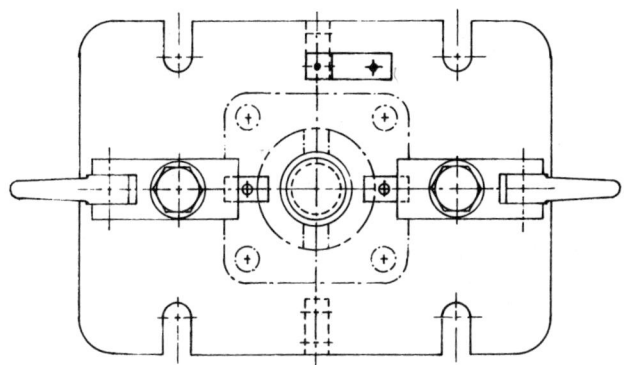

Fig. 115 Milling fixture for base casting

145

distance piece is located on its machined face by holes on location pegs. The drill bushes and screw clamp are carried on a swinging latch which is locked in position after the component has been placed in the jig. When the latch has been secured, the component can be locked in position and drilled. Fig. 114 shows a suitable fixture for holding the component during the milling operation. As the tenon is to be on the centre-line of the component, the component is held on centre by the two cut-away location pegs. Clamping is done by the inclined edge clamp holding the component up against the fixed vee. The fixture body is so designed that the resultant cutting forces are resisted by the body which is fitted with tee-slot tenons.

The base casting, shown in Fig. 115, is to have the 13 mm × 13 mm slot cut across the boss. The component is held on the large locating plug and on the tenons which fit into the 13 mm wide slots previously machined. Cam clamps hold the component on to the base plate which itself is held to the machine table by means of tee-slot tenons and bolts.

A hardened steel setting block is fitted to this fixture. It is positioned so that the fixture can be set, relative to the cutters, with the minimum of setting-up. In use, the fixture which has been bolted to the machine, is brought up to the cutters and set in relation to the cutters by inserting a feeler gauge between the setting block and the cutter. Setting blocks can be fitted to most milling fixtures. It is recommended that a feeler gauge of 0·75 mm thickness should be used in conjunction with setting blocks.

EXERCISES

1. Fig. 116 shows a brass bush. The component is machined all over except for the two holes drilled and reamed 10 mm diameter. Design a drilling jig suitable for producing the holes in large numbers of components.

2. The operation schedule for the lever, shown in Fig. 117, contains the following as the last two operations:
   Operation 4. Mill 3 mm wide slot.
   Operation 5. Drill and tap clamping screw holes.
   Design (*a*) a suitable fixture for operation 4;
   (*b*) a drilling jig for operation 5.

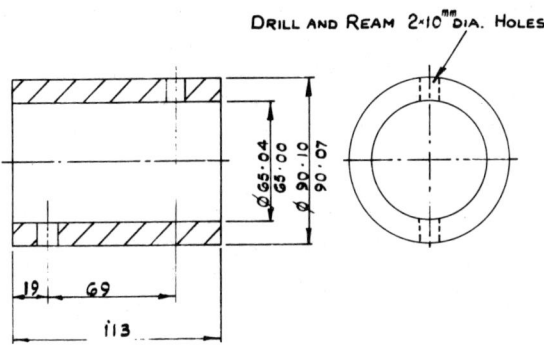

DRILL AND REAM 2×10ᵐᵐ DIA. HOLES

Fig. 116

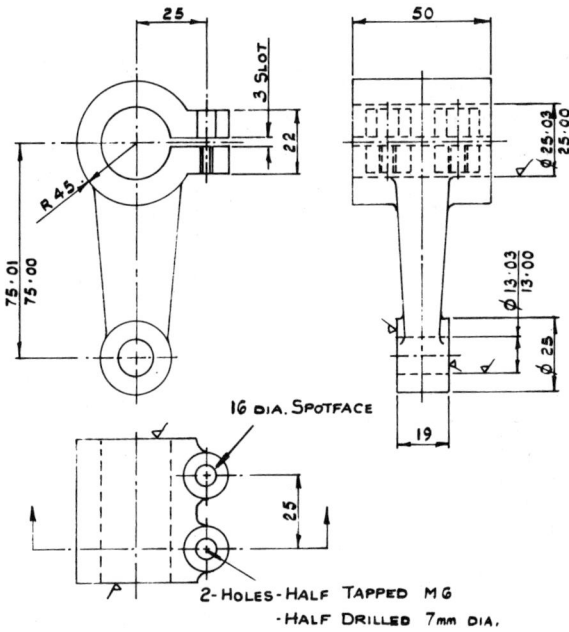

Fig. 117

3. The pivot pin, Fig. 118, is produced in large numbers on a capstan lathe from a 40 mm diameter mild steel bar. The flats are produced during a second operation by straddle milling.

Design a fixture to enable 10 of these components to be milled at one setting.

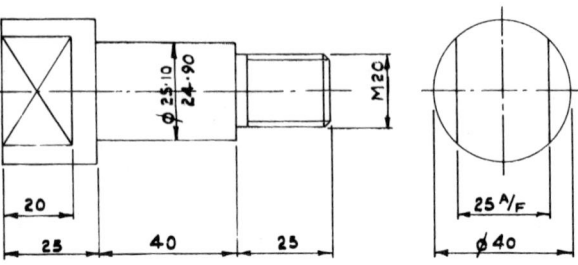

Fig. 118

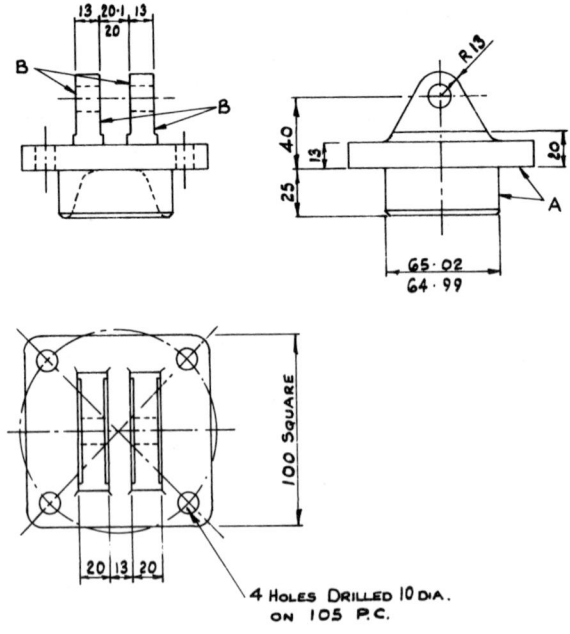

Fig. 119

4. The component shown in Fig. 119 is received as a steel forging in batches of 200.

   Operation 1 is to machine the 65 mm diameter boss and adjacent face (marked A).

   Operation 2 is to drill the 4 × 10 mm diameter holes on a single spindle drilling machine.

   Operation 3 is to mill the 4 faces of the 2 lugs (marked B) on a horizontal milling machine.

   Design (*a*) a drilling jig for operation 2;

         (*b*) a fixture suitable for operation 3.

5. Draw or sketch in good proportion, the general arrangement of a jig suitable for the drilling and tapping of the three clamping screw holes in part S of Fig. 182A in Chapter 13. First make a rough sketch assuming reasonable dimensions for the part. Give three critical dimensions on the jig.

   (C.G.L.I.)

6. Sketch or draw in good proportion, choosing views so that the construction and operation are clear, a fixture suitable for holding the bracket B in Fig. 186, Chapter 13, whilst the 3 mm slot is being cut with a slitting saw on a milling machine. All other machining may be assumed to be completed.

   Give the main dimensions necessary for the manufacture of the body of the fixture, but not for the details or the fittings.

   (C.G.L.I.)

7. Using neat sketches or drawings show the main details of a jig suitable for drilling (but not tapping) the 5 mm diameter oil hole in the housing shown in Fig. 194B, Chapter 13. All other machining of the housing may be assumed to be completed. The method of operation of the jig should be made clear. Dimensions are not required.

   (C.G.L.I.)

Mechanical Design

Briefly, design is the solution of engineering problems with respect to working principles, manufacturing method, materials and economy. Many designs are, in fact, examples of good practice which have been assembled to form a whole. In many design problems, extensive calculations of stresses, loads and proportions are not required. This is not to say that calculation is never involved in the working out of a good design; for example in the design of structural steelwork, many calculations are necessary before the finished design is complete. However, this chapter is concerned with that aspect of design involving good practice where the dimensions of the components can be estimated. The calculation is kept to a minimum and, where it is unavoidable, it has been limited to the experience of the reader. Indeed, many of the problems at the end of this chapter can be solved without any calculation whatsoever.

Manufacturing companies whose products are widely used, produce excellent catalogues and technical information, and it is suggested that wherever possible, the reader should examine this material.

The remainder of this chapter will attempt to cover as many aspects of good practice as possible.

Bearings

Wherever moving parts come into contact with each other or with a stationary surface, surface friction will tend to oppose the movement between them. Heat may be generated and efficiency will be lost unless some attempt is made to reduce the friction to a minimum. When two adjacent parts are in contact and move relative to each other, the provision of a well designed bearing will greatly improve the efficiency of the movement. Most bearings comprise a rotating shaft supported in a hole and it is essential that the following factors be considered when a suitable bearing is being designed:

(*a*) load on the bearing (type and magnitude);
(*b*) rubbing speed of bearing (peripheral speed of shaft);
(*c*) provision of lubrication;
(*d*) shaft diameter;
(*e*) length of bearing;
(*f*) considerations of adjustments and renewals.

It is only when these items have been considered, that the choice of bearing can be made.

Plain Bearings

When a shaft revolves intermittently or at low speeds with light loads, it can be supported in a plain bearing shown at Fig. 120. This is known as a *plain journal bearing*.

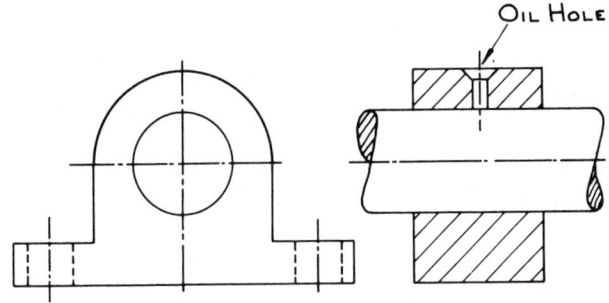

Fig. 120 Plain journal bearing

It is made from grey cast iron and is to some extent, self-lubricating. However, if lubrication is required, an oil hole can be provided. There is a severe limitation to the use of such a bearing, namely, that when the bore becomes worn, it is necessary to replace the whole bearing. This is a costly operation and in order to overcome it, the cast iron bearing becomes a housing to support a bush as in Fig. 121. On small diameters, the bush is made a press fit into the housing and no other locking device is needed to prevent the bush rotating. On larger diameters, some method of holding the bush in position is required. Fig. 122A shows a bush held in its housing by means of a grub screw, tapped half in the bush and half in the housing. Fig. 122B shows a flanged bush prevented from rotating by a peg which engages in a slot cut in the housing. Another reason for using a bush is because of the high cost of some bearing metals; if the bearing was made entirely of the bearing metal its cost would be exorbitant, in addition, the bearing metal would probably not be satisfactory as a housing.

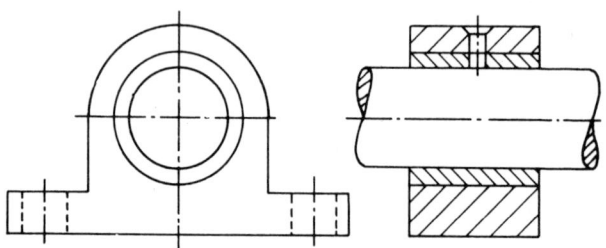

Fig. 121 Plain bearing with bush

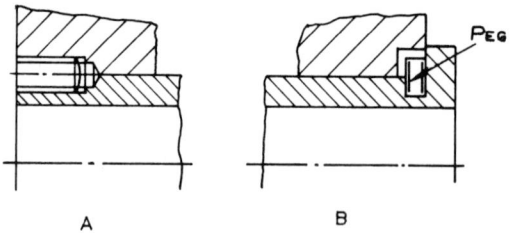

Fig. 122 Methods of securing bearings

In the case of the bush, its life is finished when it becomes worn and no adjustment can be made, but it can be replaced at moderate cost. The split bush and housing shown in Fig. 123 can be adjusted to take up wear in the bush. This type has the advantage that, being split, adjustments and replacements can be made without dismantling the full length of shaft. Again there are various methods of preventing the split bush from rotating in its housing, and these are shown in Fig. 124. At A, a cast projection on top of the bush engages in a recess in the bearing cap, the projection carrying the

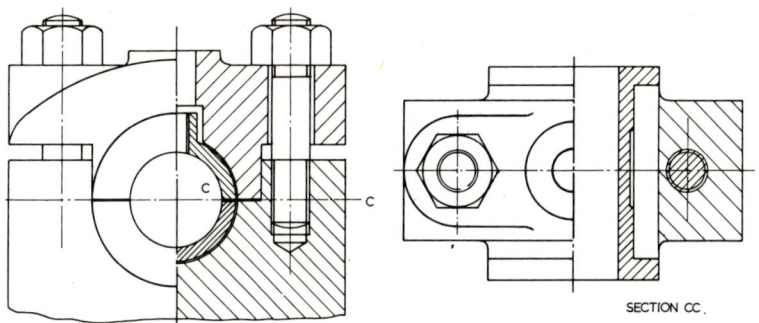

Fig. 123 Split bush and housing

SECTION CC

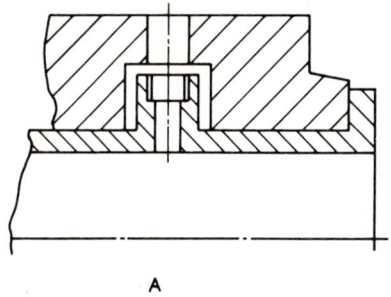

A

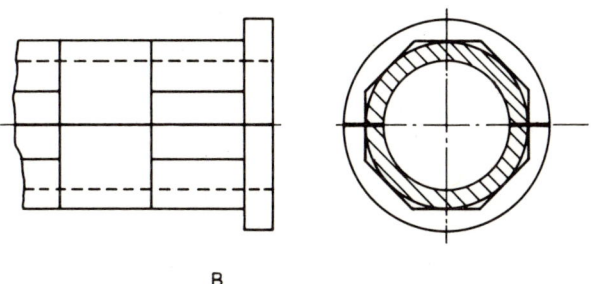

B

Fig. 124 Methods of preventing split bushes from rotating

153

oil way in to the bearing. At B, the split bush has an octagonal shape which mates with a similar shape in the housing.

The bearings so far discussed have been intended for use where the loading on the shaft is in a radial direction only. When the loading is axial, some form of thrust bearing will be required.

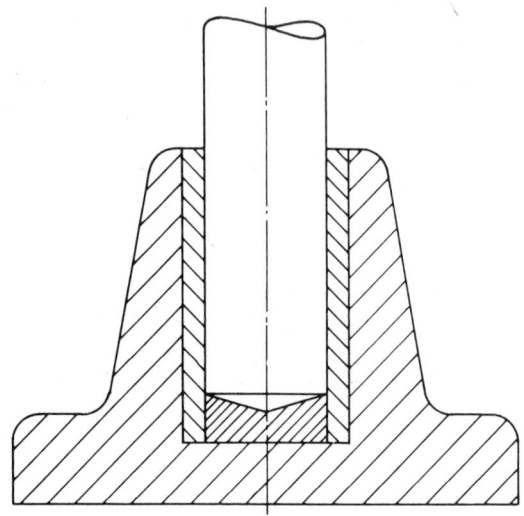

Fig. 125 Vertical thrust footstep bearing

Fig. 125 shows a thrust bearing intended for use at the end of a vertical shaft. This is often called a footstep bearing. There are many occasions when the loading on a bearing is a combination of radial and axial loads. This occurs in the bearings supporting single helical gears and worms. The bearings of the propellor shaft of a ship are also loaded axially and radially.

The thrust bearings shown at Fig. 126 can be used for radial loading and axial loading in one direction only. Where axial loading takes place in both directions, two collars are required on the shaft to resist the thrust. The schematic layouts at Fig. 127 show the placing of the flanged bearings with respect to the collars. The choice of bearing material is dependent upon the bearing pressure to which it is subjected. Bearing pressure is the load per unit of projected area of the bearing; the projected area means the area

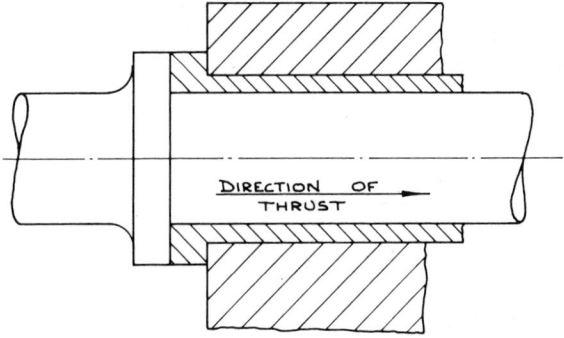

Fig. 126 Typical thrust bearing for thrust in one direction only

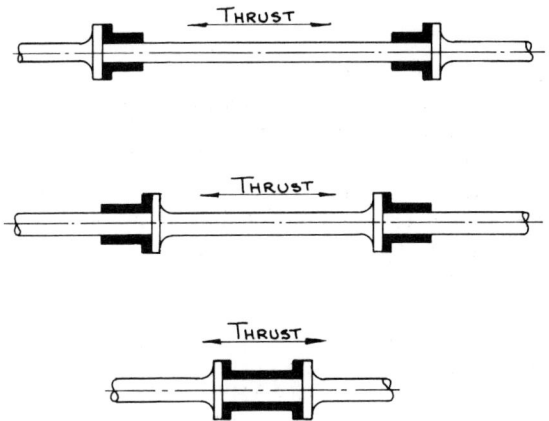

Fig. 127 Alternative arrangements of thrust collars

projected on a plane at right angles to the direction of the load. Fig. 128 shows the projected areas of journal and thrust bearings.

$$\text{Intensity of bearing pressure} = \frac{\text{total load on bearing}}{\text{projected area of bearing.}}$$

It will be appreciated that the pressure changes according to the dimensions of the bearing. Pressure can be reduced by increasing the length of the bearing and/or the diameter of the shaft (the converse is also true). The shaft diameter is usually chosen first as this

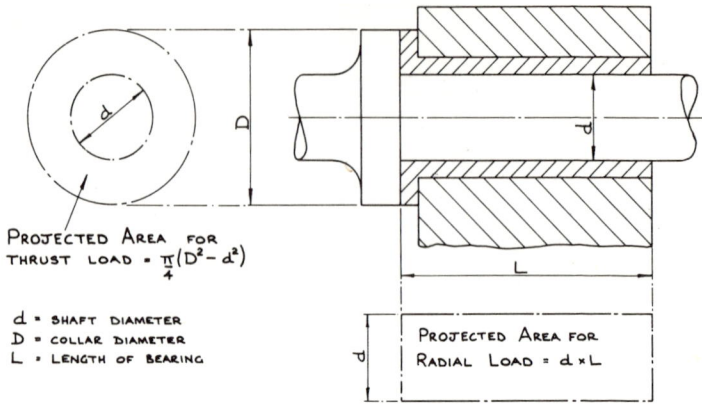

PROJECTED AREA FOR
THRUST LOAD = $\frac{\pi}{4}(D^2 - d^2)$

d = SHAFT DIAMETER
D = COLLAR DIAMETER
L = LENGTH OF BEARING

PROJECTED AREA FOR
RADIAL LOAD = d × L

Fig. 128 Projected areas

depends on the strength of shaft required. The possible length of bearing may be limited and this is another factor to be taken into consideration. However, the length of the bearing is selected in conjunction with the maximum pressure that can be supported by the bearing material chosen.

The table gives average values of maximum allowable pressure for some materials together with maximum speeds.

Material	Max. pressure	Max. speed
Lead base alloys	12 MPa	600 m/min
Tin base alloys	12 MPa	600 m/min
Bronze	18·5 MPa	275 m/min
Copper-lead	21 MPa	900 m/min
Aluminium-tin	27·5 MPa	750 m/min
Cast iron	3·5 MPa	40 m/min

Bushes for bearings are commonly made from gun-metal or phosphor-bronze, but steel shells supporting white metal and plastic bearing materials are also available. Chapter 3 discusses the various bearing metals available.

Journal bearings, still in common use, are made by casting white metal (tin based bearing alloys) into steel and cast iron shells (Fig. 129). The bearing metal is melted, cast into the shell round a central core and when cool, the bearing in its shell is machined to size. The bearing metal must be keyed into place in the shell,

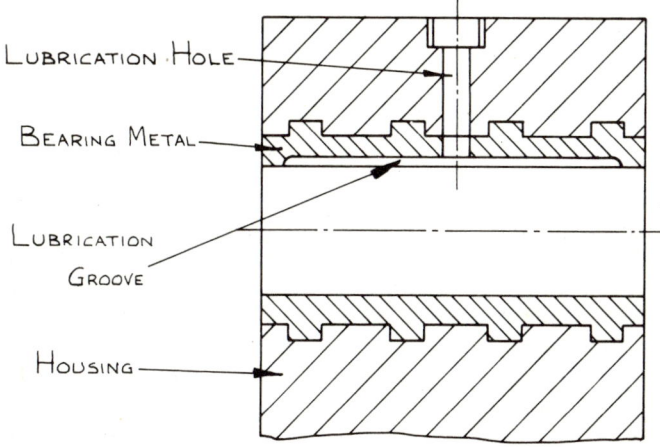

LUBRICATION HOLE

BEARING METAL

LUBRICATION
GROOVE

HOUSING

Fig. 129 White metal bearing

therefore recesses are machined in this shell for that purpose.

All bearings require lubrication to reduce friction and some means of supplying oil or grease to the bearing should be provided. Holes which pass from the outside of the housing to the rubbing surfaces are the simplest to produce. An oil-can may be used from time to time or the oil hole can be modified to take a wick, a drip feed oiler, a grease nipple or an applicator. As some means of spreading the lubricant inside the bush is also required it is best to provide an oil groove which runs parallel to the shaft and almost to the ends of the bush, as in Fig. 129.

The advent of sintered metal oil-impregnated bushes has made the lubrication of small lightly loaded bushed bearings unnecessary. These bearing bushes are 'self-lubricating' in so far as their lubricant supply usually lasts the life of the bush. If a supplementary lubricant supply is required, it can be provided by drilling a hole in the bearing housing and fitting the hole with a suitable cover. A hole through the bush and grooves are not required because the bearing is porous. The hole acts as a reservoir which can be filled with oil or oil-soaked felt. The oil will then permeate through the bearing as it is needed. Sintered metal bearings are used in intricate and compact assemblies and where lubrication is likely to be difficult or neglected.

Antifriction Bearings

Whilst well designed plain bearings still find many uses, they are quite inefficient by comparison with the high precision ball and roller bearings. All varieties of antifriction bearing substitute rolling friction for the sliding friction found in plain bearings, consequently, less power is required to overcome frictional resistance.

Antifriction bearings are made in inch and metric sizes. The inch sizes are interchangeable with inch size bearings from all manufacturers in this country and the metric sizes are interchangeable with those manufactured on the continent and in the U.S.A.

Many varieties of bearing are available and the selection and application of the correct bearing to any particular job depends on several factors:

(*a*) Is the bearing suitable for carrying axial and radial loads, and in what proportions?

(*b*) Is it suitable for the speed range of the application?

(*c*) Is it subjected to shock or vibratory loads?

(*d*) Will it cope with small errors in alignment?

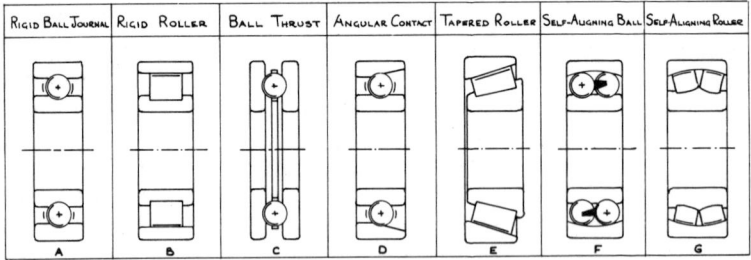

Fig. 130 Typical antifriction bearings

Fig. 130 shows the more common varieties of antifriction bearing available but there are many other types. A rigid ball journal bearing is shown at A, and this is probably in greater use than any other type. It will carry radial and axial loads with combinations of both, and finds many uses in industry in such applications as gear boxes, roller conveyors, mechanical handling equipment, electric motor spindles and plummer blocks for shafting; it is capable of locating a shaft in both directions. At B a rigid roller bearing is shown; this type can carry very high radial loads and is suitable for shock or vibratory loads. Roller bearings cannot take axial loads nor can they locate a shaft as the design of the bearing permits free lateral movement of the component parts. This characteristic can be used to advantage when a shaft is located by a rigid ball bearing at one end and supported by a roller bearing at the other. The lateral movement will allow the shaft to expand without affecting the performance of the bearings. The inner track of the bearing shown is lipped to guide the rollers but various combinations of lips can be obtained; the inner track could be left without lips whilst the outer track is lipped. Often both tracks are lipped and sometimes the assembled bearing has three lips in any position.

The high radial load capacity of the roller bearing makes it useful in applications where heavy loads are applied to the shaft as in drive shafts for conveyor belts, bucket elevators, gear boxes and other heavily loaded service conditions.

The ball thrust bearing shown at C will not take any radial load and is designed to carry axial loads only. Modern practice is tending to discard the use of this type of bearing as they can only be used at low and medium speeds. For high speeds with high axial loading, a double purpose ball bearing is preferred. However, they still find use in applications where vertical shafts are used. Because most shafts require a bearing to keep the shaft in correct alignment, another bearing of some form must be used in conjunction with the ball thrust bearing. In a worm gear or single helical gear box, this support is usually given by a ball or roller bearing, but for many applications a plain bearing will suffice.

The axial load keeps the bearing parts in close contact and for this reason they can only accept axial loading in one direction. If axial loads are expected in both directions, two thrust bearings must be used.

At D, an angular contact bearing is shown; this is truly a double purpose bearing, capable of taking thrust in one direction only or combined radial and thrust loads. Ideally, two such bearings should be fitted to the shaft, set in opposite directions. As it is the axial load which keeps the balls in their tracks, the two bearings must be accurately adjusted to eliminate end play. Its axial loading capacity at high speeds is far in excess of that of an ordinary thrust bearing. It finds use when accuracy is required such as machine tool spindles.

At E, a tapered roller bearing is shown, its purpose is similar to that of the angular contact bearing, but it will take a higher load. As this bearing will only take thrust in one direction, another, similar bearing must be used in conjunction with it. Again, all end play must be eliminated and some degree of pre-load is acceptable when it is used for accurate assemblies. This type of bearing is used in the bearings of lathe headstocks, on heavy rotating machinery such as welding manipulators and rolling mill drives.

In all applications of the bearings so far discussed, accurate alignment of shaft and housing is essential, but when this cannot be achieved or when the shaft is likely to deflect, a self-aligning bearing should be chosen. At F, a self-aligning ball bearing is shown; it should not be used for high speed applications or when axial loading is present. However, because it is self-aligning it finds many applications as a bearing for long lengths of rotating shafts. A self-aligning roller bearing is shown at G; its purpose is similar to that of F but heavier loads are possible.

The Makers' Dot Marking System

British bearing manufacturers follow a standard practice in marking the outer rings of their ball and roller journal bearings with dot marks viz: '0', '00' and '000'. These are referred to as 'one dot', 'two dot' and 'three dot' fit bearings.

This indicates the degree of diametral clearance between the balls or rollers and their tracks. The 'one dot' fit bearing, has the least clearance of all and should be used only for radial loading of the bearing or where radial and axial movement of the shaft must be restricted. Tight mounting of the tracks and high temperature applications should be avoided.

For the majority of uses, the 'two dot' fit bearing is suitable. This is accepted as 'normal' internal clearance and is used where high radial loads and light axial loading is required.

The 'three dot' bearing has the most diametral clearance and is used when a great amount of thrust load is applied to a rigid ball bearing. It is also of value when tight mounting fits are required and when the bearing is to be used in high temperature applications. A 'four dot' fit bearing is also manufactured, but its regular use is not recommended. It is used for very high temperature applications and when very heavy mounting fits are required.

The Mounting of Antifriction Bearings

There are several basic principles which must be applied when

160

designing the housings and seatings of antifriction bearings:

(*a*) in addition to supporting the shaft radially, the bearings must locate the shaft axially;

(*b*) no rigidly coupled shaft must be located at more than one point in either direction;

(*c*) allowance must be made for axial movement due to expansion or contraction of the shaft;

(*d*) rigid ball and roller bearings must be mounted accurately in alignment and be square, otherwise self-aligning bearings must be used;

(*e*) housings must be suitably sealed to protect the bearings and retain the lubricant.

Fig. 131 shows bearings in housings with the basic principles applied. Assume that A and B are mountings at either end of a shaft; at A a rigid ball journal is used to locate the shaft in both directions which is reduced in diameter to form a seating for the bearing. The screw thread is slotted to accept the tongue of a tab

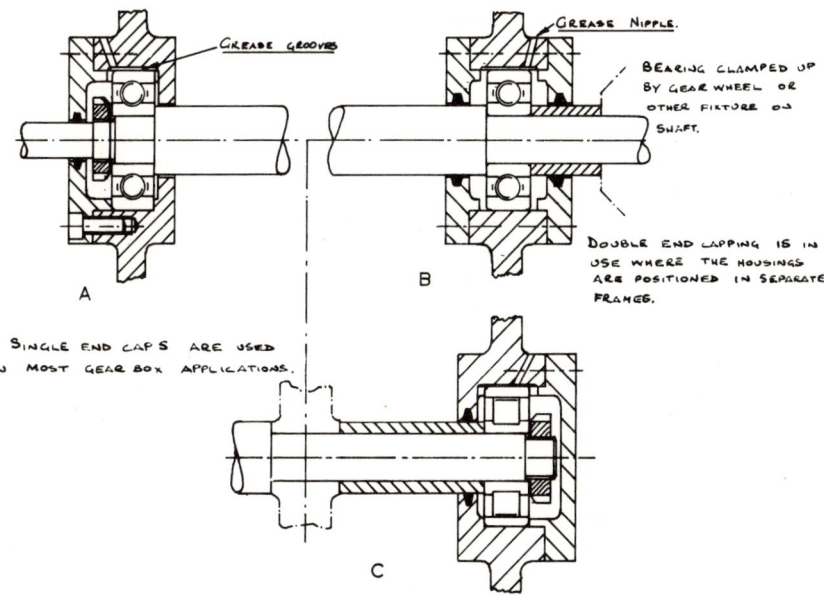

Fig. 131 Methods of mounting antifriction bearings

washer, and the locknut secures the bearing on its seating up against the shoulder on the shaft; a tab on the tab washer is bent over into a slot in the locknut to secure the whole. The housing is bored out to accept the outside diameter of the bearing but care must be taken to ensure that the back of the housing is relieved so as not to rub on the inner track of the bearing. This type of housing is used where the shaft is to be totally enclosed, as in a gear box. The end cap is made to fit the housing and hold the outer track firmly against the back of the housing when the screws are tightened down. If the located end of the shaft is to be driven or connected to some other shaft, it must pass through the end cap and be provided with a suitable seal. The bearing must be lubricated and grooves should be cut in the housing to allow the lubricant to circulate.

At B, a rigid ball journal is used as a non-locating bearing, which is permitted to slide in the housing and the end caps do not hold the bearing firmly. This type of housing and end capping is used when the housings are positioned in separate frames or otherwise where the shaft is not enclosed. The inner track is clamped against the shoulder on the shaft by means of a sleeve which in turn is held in place by a gear wheel or pulley. Assume that end B is replaced by end C: a rigid roller bearing is now being used as a non-locating bearing. The outer track is fixed in the housing by a fully closed end cap, yet variations in shaft length can be accommodated by the bearing itself. The inner track and rollers can move laterally in the outer track without affecting the load carrying capacity of the bearing. In this case, a pulley or gear wheel is positioned by means of a sleeve which is held in place by the bearing with locknut and tab washer. Any combination of end capping, housing arrangement, sleeving and locking systems can be employed, provided that the basic principles are followed.

Housings and Shafts

Housings should be rigid in design and should preferably not be split where it can be avoided. When the bearing centres are reasonably close together the housings should be made in one piece so as to enable them to be more easily machined in alignment. The housings should be bored to a good smooth finish to the tolerances suggested by the manufacturer.

It is recommended that aluminium, or other light alloy housings should not be used because:

(a) differential expansion of housing and bearing loosens the fit of the bearing track in its housing;

(b) the housing bore is likely to deform and increase in diameter due to shock loading;

(c) to prevent the effects of (a) and (b), the outer track must be a good interference fit in the housing. This is not good practice as lateral movement of the outer track is prevented.

The bearing seat on a shaft should be finished by grinding but if this is not possible a finish similar to that of the housing should be given by smooth turning. The tolerance which is given to the shaft seating should be extracted from the manufacturer's catalogue. It is not necessary to harden the seatings on the shaft.

Lubrication of Antifriction Bearings

The purpose of a lubricant is primarily to reduce the friction in a bearing assembly but it can also be used as a coolant on the bearing and as a seal to prevent dirt and grit from entering the bearings. There are two forms of lubricant, namely, grease and oil. Oil, a very reliable lubricant, must be used at high speeds, but grease is perhaps easier to apply and is more convenient.

(a) *Grease.* When low speeds are encountered, and where the temperature does not exceed 120°C, grease can be used extensively. Under normal conditions, regreasing at 6-monthly intervals is adequate but the housing should be tightly packed with grease. A pressure grease gun should be used to administer the grease which should be pumped in until it is seen to seep out past the sealing arrangement. The grease gun should be used at frequent intervals.

The tightly packed grease tends to prevent foreign matter entering the housing, but the charging of the housing until the grease seeps out, will carry away any grit which may have begun to get past the seals.

Any acid-free grease is suitable for most applications, but when wet conditions exist, a lime-base grease should be used to prevent emulsification of the grease.

Acceptable maximum speeds for grease lubricated bearings are difficult to estimate. As the speed of the shaft increases the question of oil or grease lubrication becomes more complex. However, bearing manufacturers give recommended maximum speeds for grease lubrication in their catalogues and these should be consulted for any specific design. As a general guide the following maximum speeds are given:

Bore diameter of bearing	*Maximum speed for grease lubrication*
up to 25 mm	6 000 rev/min
25 mm to 50 mm	4 000 rev/min
50 mm to 75 mm	2 000 rev/min
75 mm to 100 mm	1 000 rev/min

Where speeds are in excess of these figures oil lubrication should be used.

(*b*) *Oil.* Good quality mineral oil should always be used for the lubrication of bearings and it can be applied to the bearing in many different ways. The drip feed and wick feed applications find many uses and the newer oil mist system is being widely adopted. In this system, ideal for high speeds, the oil is atomized and blown into the bearing by compressed air.

A common method of lubricating bearings in a gear box is by splash feed, but the oil must be suitable for both gears and bearings. The bearings are situated near the gears and are lubricated by splash as the gears enter and leave the oil reservoir. Careful design is required to ensure efficient lubrication.

Another simple method of oil lubrication, not suitable for high speeds, is that of filling the housing with oil to a depth halfway up the lowest ball. To prevent overfilling, a hole in the housing should be drilled to allow for seepage, alternatively, an external oil level sighting tube could be used.

The ultimate lubrication system for high speed bearings is an oil circulation system using a forced feed. An oil pump is usually adopted and this can supply a number of bearings with oil, but a cooling unit is required to prevent the oil from becoming overheated.

Seals and Glands

A seal can be defined as a device for containing lubricant in a bearing whilst the shaft revolves, but it also prevents ingress of foreign matter. A gland is also a seal in one aspect as it prevents foreign matter from entering a housing but its main function is to prevent the leakage of fluid around a shaft whilst the shaft revolves or reciprocates. The seal is used primarily in connection with bearing arrangements on gear boxes and industrial assemblies involving bearings. The gland is often used to prevent leakage on pumps, steam cylinders and stop cocks.

Seals. The type of seal selected must suit the application, i.e. a seal suitable for grease lubrication may not be suitable for sealing against oil. Fig. 132 shows an end cap having grease grooves machined in it; the grease fills the grooves and makes an effective seal. The shaft must be a fine running fit in the bore of the cap and ovality or eccentricity of shaft and bore must be eliminated. A diametral clearance of 0·25 m to 0·35 m between shaft and bore is

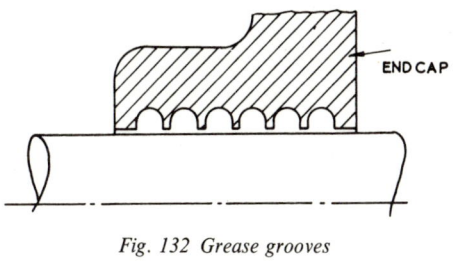

Fig. 132 Grease grooves

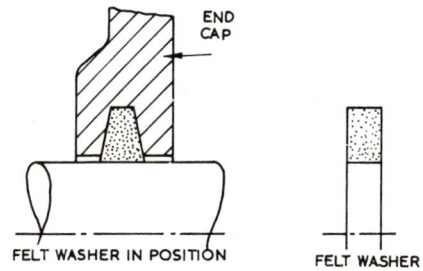

Fig. 133 Felt washer

considered acceptable. Fig. 133 shows the use of a felt washer suitable for sealing against grease at low and moderate speeds. The felt tends to harden and become impregnated with grit and this affects the efficiency of the seal. Whilst the seal contacts the shaft, a clearance of 0·5 m is usually present between the shaft and the bore of the cap. The felt washer can be fitted singly or in pairs depending upon the efficiency required of the seal, and it is usual to soak the felt washer in oil before use.

Labyrinth seals are effectively used with both grease and oil lubrication but with oil, the shaft should be in a horizontal position. Fig. 134 shows a labyrinth seal. It must be accurately machined and fitted without eccentricity or ovality otherwise centrifugal pumping action occurs causing the lubricant to be drained from the bearing.

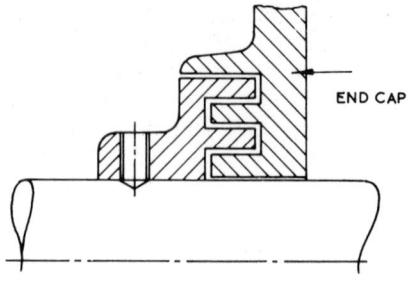

END CAP

Fig. 134 Labyrinth seal

Radial clearances between labyrinth faces should be 0·1 mm whilst axial clearance should be 0·5 mm at the located end of the shaft and 1 mm at the non-located end.

A large variety of commercially obtainable seals are available from various manufacturers. Fig. 135 shows a section through a typical propriety seal suitable for grease and oil, where the pressure is under 0·5 MPa. It is usually made of synthetic rubber or plastic material which is bonded to a metal supporting ring to facilitate fitting. The area of contact with the shaft is reduced to a minimum and a garter spring supplies light uniform pressure. The shaft should be ground and polished to 5 microns to obtain a satisfactory result with ovality and eccentricity reduced to a minimum. Fig. 136 shows methods of mounting these seals. At A, the normal method of mounting is shown whilst at B, the normal end cap is machined to accept the seal. Hardening of the shaft is not recommended and mild steel is thus a suitable material.

A recent innovation in the form of a V-ring (Fig. 137), has several marked advantages over ·the conventional seal. It has no metal parts and is made from nitrile rubber. The makers claim that special machining is not required to fit it to an assembly, it will accept ovality and eccentricity without affecting its function and a seal can still be achieved even if the shaft is not in alignment. Its only requirement is that the face against which the seal is to rub, should be finished by fine turning, followed by polishing with emery cloth. A bearing seal with V-rings is shown in Fig. 138.

The seals previously discussed are all external to the bearing. Rigid ball bearings are available with built in seals charged with grease, and whilst these are extremely suitable for some applications, the bearing life is limited to the life of the grease charge, which is about 10 000 hours. The bearing cannot be dismantled and recharged with grease.

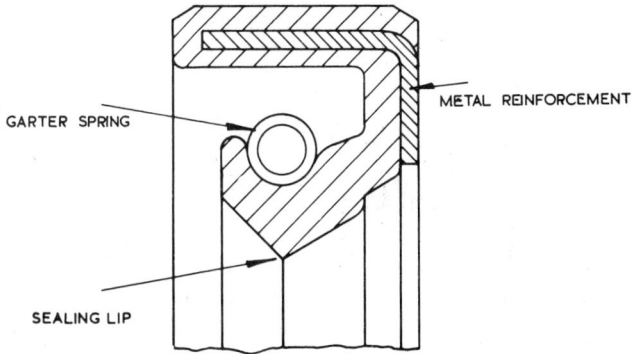

GARTER SPRING

METAL REINFORCEMENT

SEALING LIP

Fig. 135 Section through oil seal

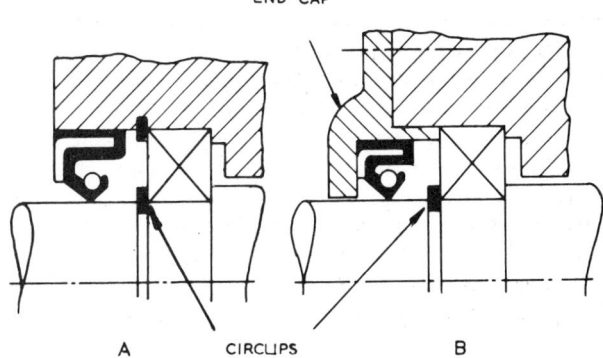

END CAP

A   CIRCLIPS   B

Fig. 136 Methods of mounting seals

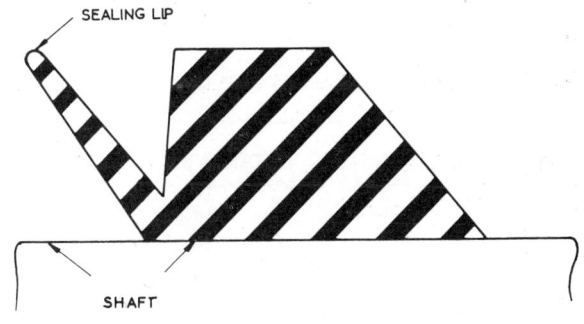

SEALING LIP

SHAFT

Fig. 137 V-ring oil seal

167

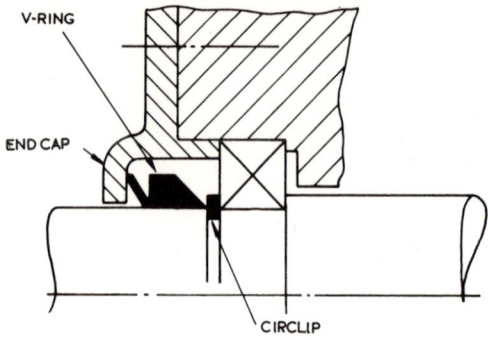

Fig. 138 Use of the V-ring

Glands. Fig. 139 shows an assembly where some form of packing is compressed into a stuffing box around a reciprocating or revolving shaft by means of a gland. There are many names for this assembly but the writers prefer to call it a gland, giving each component its individual name. The gland is used where a shaft passes through the casing of a vessel containing fluid. It might be a water pump where the shaft is brought through the casing to be driven by a motor or a steam engine piston. The arrangement is usually found on gear pumps, valves and cocks, and its purpose is to allow the shaft to move freely yet prevent the escape of fluid from the vessel. The stuffing box can be integral with the vessel or can be bolted on separately, but its function is the same, i.e. to support the rod in a neck bush and to carry the packing and the gland cap.

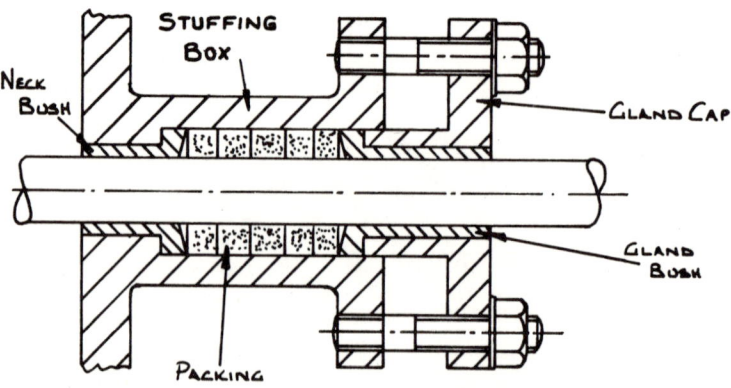

Fig. 139 Stuffing boxes, packings and glands

The gland cap also carries a bush, called a gland bush. If the shaft has a short overhang the phosphor-bronze bush may be omitted. The packing is usually made from woven asbestos in the form of a square cord and impregnated with graphite for lubrication purposes; gland packings can also be made from canvas and rubber impregnated with graphite or white lead. To form a seal, the gland cap is clamped down on to the packing by means of nuts and studs. The packing is thus compressed in the stuffing box and on to the shaft, so preventing fluid losses.

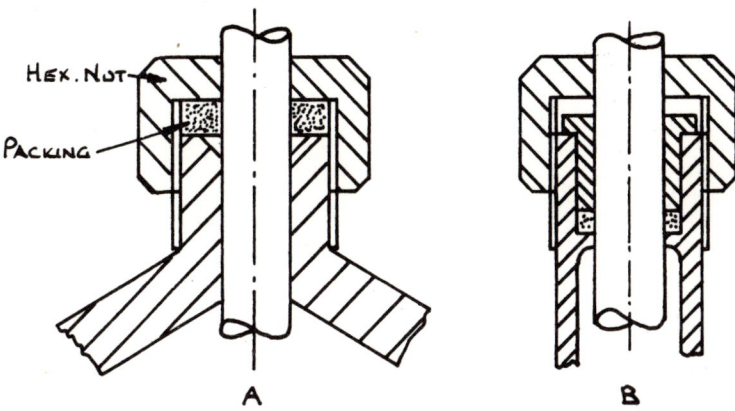

Fig. 140 Screwed glands

Simple screwed glands are shown in Fig. 140 and these are often used on small valves. At A, the packing is compressed by the hexagon nut whilst at B, the nut applies pressure to a sleeve which in turn compresses the packing. In this type of gland, the packing can be a winding of asbestos string or a synthetic rubber washer or a ring having a circular cross section, called an 'O' ring.

Bosses and shafts

One of the most common problems in engineering design is that of fastening a bossed component such as a gear or pulley on to a shaft, so that it cannot revolve or move axially.

Fig. 141 shows four common methods of fastening bosses to the centres of shafts. The arrangement shown at A involves the use of one or two grub screws, with conical points, and they locate in

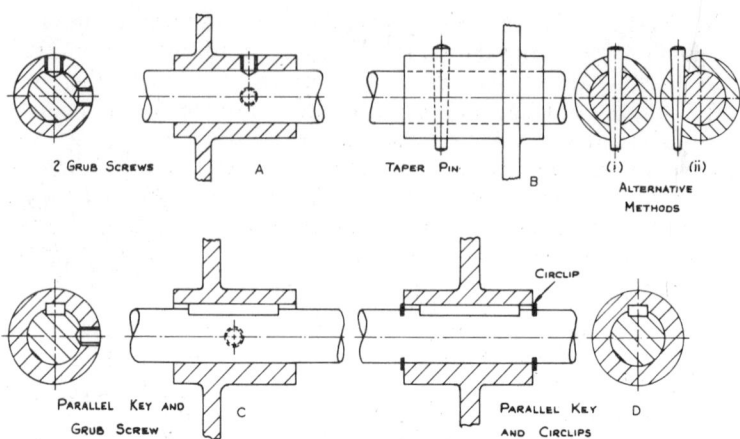

Fig. 141 Methods of fastening bosses to shafts

'dimples' in the shaft. This method is ideal if light loads and powers have to be transmitted but there is always a danger that an overload will cause the boss to turn with the result that the screws score the shaft. Vibrations may also cause the screws to slacken and the boss to work loose.

The use of a taper pin, as at **B**, is a more positive arrangement. The section shows alternative methods of using the taper pin. At (i), the pin is driven through the centre of the shaft but this weakens the shaft slightly. To allow maximum loading to be transmitted by a taper pin the method shown at (ii) should be adopted. The taper pin arrangement at (i) has use as a safety measure. By the selection of a pin of suitable cross section and material, the torque at which the pin will shear can be calculated. This finds extensive use in the machine tool industry on lead and feed screw drives.

A gib head tapered key can be used for the purpose of securing a boss axially and radially but a long keyway and ample assembly space will be required. A parallel key can overcome the problem of rotation, but axial location must still be provided. If the axial forces are not too large, then a conical pointed grub screw can be used as at **C**. At **D**, the same parallel key is used and axial location is provided by circlips in grooves machined in the shaft.

Circlips are being widely used for location purposes, they can be used externally on shafts or internally in bores and their scope is limited only by the ingenuity of the designer. A few suggested uses

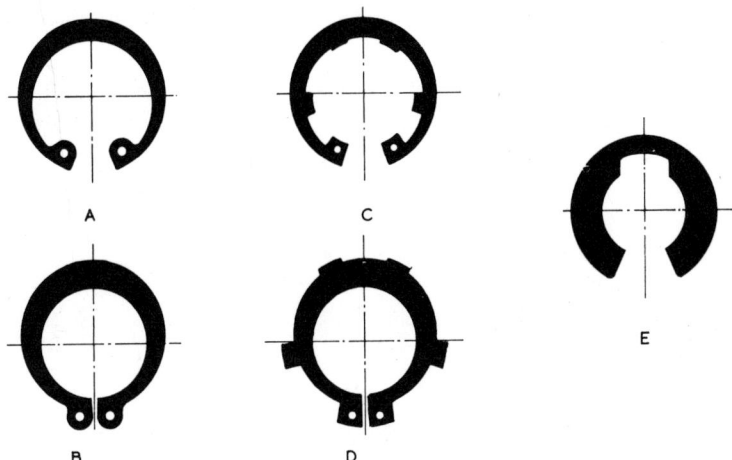

Fig. 142 Circlips

include the location of bearings in housings and on shafts, shaft locations when used with plain bearings, location of bushes and sleeves in bores and, of course, the location of the gudgeon pin in internal combustion engines. Fig. 142 shows internal and external circlips made from spring steel; the holes are provided for use with special circlip pliers which draw in, or push out the ends of the circlip so as to enable it to be mounted in its groove.

The one shown at A is a general purpose circlip for use in bores and is classed as an internal circlip. For use on shafts and spindles, the one shown at B is used and is classed as an external circlip.

The clips at C and D are for use in retaining ball and roller bearings in their locations, the lugs on the circlips prevent the inner and outer tracks of the bearings from moving. C is an internal circlip whilst D is for use externally.

A spring clip is shown at E; this is not mounted like a circlip but is pushed into a groove from the side whereas a circlip is sprung and mounted by pushing it axially along a shaft or bore into its groove. It is, however, used successfully for location purposes.

Dimensions of circlips and circlip grooves can be obtained from the manufacturers' catalogues.

Fig. 143 shows some methods of mounting a bossed component on to the end of a shaft so that it will not rotate, or move axially; at A, the shaft is made square and fits a square hole in the boss. It is

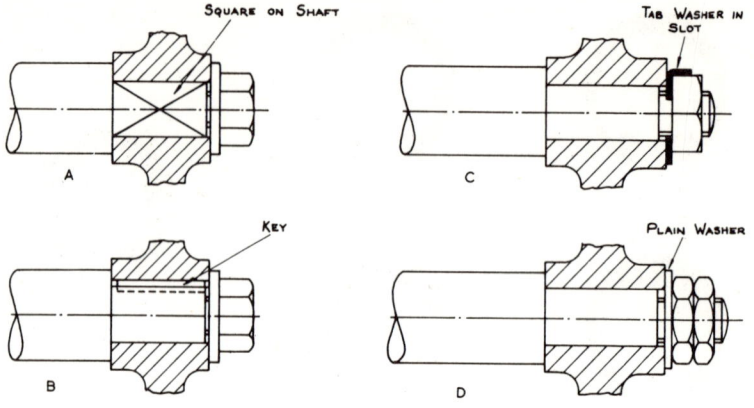

Fig. 143 *Methods of mounting bosses on ends of shafts*

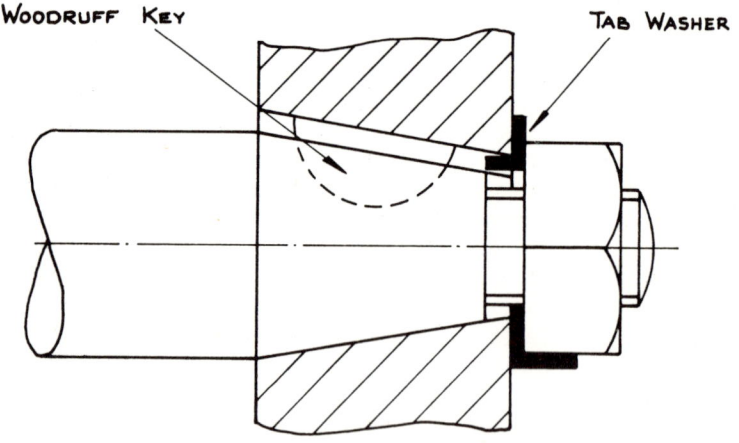

Fig. 144 *The use of a tapered shaft*

prevented from coming off the shaft by a washer and set screw. This system is ideal for use where the bossed component is a handle and accuracy is not so important. The major problem in manufacturing this arrangement, for a component which must run true, is to ensure that the square and the square hole are truly axial; this can be achieved but the cost is high. A much simpler method of ensuring concentricity is shown at B: the shaft is reduced in diameter to form a shoulder and a keyway with key is provided. The shaft could also be splined to prevent rotation with the locking arrangement similar to that shown at A. In both these cases, the shaft is slightly shorter than the boss thickness, so that the washer, under the tightening action of the set screw, grips the boss tightly against the shoulder provided on the shaft.

In the last example, location could have been provided by a circlip, whereas a tab washer and locknut, at C, or two locknuts and a washer, at D, could provide an alternative in either case.

The use of a tapered shaft is shown in Fig. 144: the locknut draws the two tapers together to provide an excellent fit which does not allow axial movement of the boss whilst the Woodruff key provides additional security for transmission of rotation. This particular arrangement is widely used for fastening flywheels to motor cycle crank shafts.

Conditions are only slightly different when a bossed component is required to be located axially but to be free to rotate. In Fig. 141D, if the key is removed, and the shaft left plain, the desired conditions could be obtained, but the efficiency of the assembly would be low. A bearing is now present and it should be designed as such. One method is to fit a bush of some low friction material into the boss. The bossed component must be free to rotate and a small amount of end play must be tolerated. The use of circlips is also in question in an application such as this, although the use of spring clips (Fig. 142E) may be acceptable. In order to locate the revolving boss, a better idea is to use loose collars which lock into place on the shaft by means of cone-ended grub screws; such an assembly is shown in Fig. 145. If considerable end thrust is expected, washers

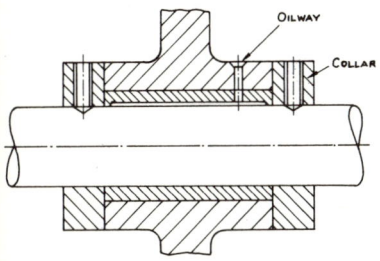

Fig. 145 Bushed rotating boss

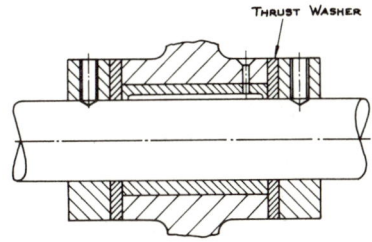

Fig. 146 Use of thrust washers

173

of bearing material can be inserted between the collars and the bosses as in Fig. 146. If the bush can be made flanged, only one washer will be required. When the boss is long, it is customary to bush both ends of the boss as in Fig. 147, to save material and machining time; the bush flanges acting as thrust washers in this case.

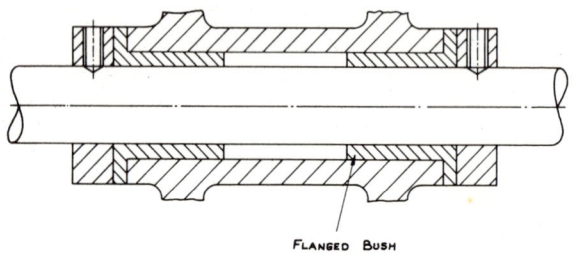

FLANGED BUSH

Fig. 147 Flanged bushes in a long boss

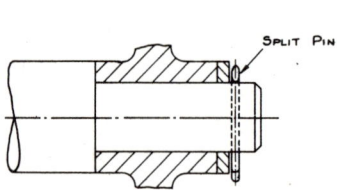

SPLIT PIN

Fig. 148 Rotating boss on end of shaft

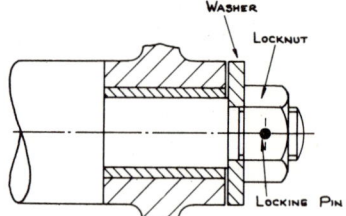

WASHER

LOCKNUT

LOCKING PIN

Fig. 149 Bushed boss with locknut and washer

Fig. 148 shows a simple arrangement when a boss is to rotate at the end of a shaft without axial movement, this simple but effective method employs a stepped shaft to locate the boss in one direction and a washer and split pin in the other. A more sophisticated method of doing the same job is to secure the washer up against another shoulder by means of a nut, Fig. 149. When the rotating boss is to be flush with the end of the shaft, an arrangement such as that shown in Fig. 150, may be used to advantage.

We have considered the fitting of bosses on to shafts but the reverse problem of fitting shafts into bosses is, in many ways, similar. In the latter case the boss supports the shaft, which is

174

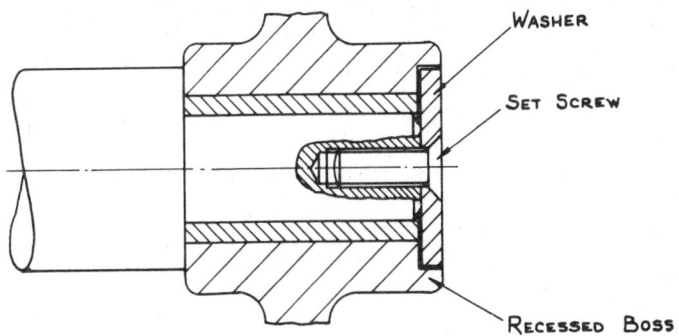

Fig. 150 Flush end rotating boss

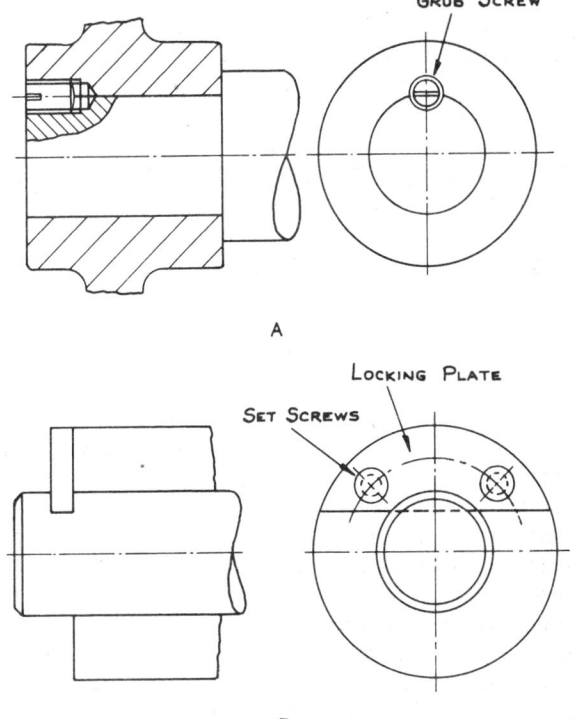

Fig. 151 Methods of securing shafts in fixed bosses

175

either held or free to move, whilst the boss remains stationary, fixed in the structure of the machine to which it belongs.

The methods shown in Fig. 141A and B are quite suitable when a shaft is to be fixed into a boss and can be adapted to suit a fixing at the end of a shaft in addition to the condition shown. The shaft is also located in both directions and need not be fixed in its boss at the other end.

Keyways are relatively difficult to machine in a fixed boss and the methods shown at Fig. 141C and D are not usually adopted.

Fig. 151 shows two other methods of fixing the end of a shaft in a fixed boss; at A a grub-screw is tapped half and half in the boss and shaft, whilst a locking plate which fits into a slot cut in the shaft is shown at B. The locking plate is held fast by two set screws.

When the shaft is to rotate in the fixed boss, a bearing is again present so that bushes and collars must be adopted to give adequate support and location.

Many of the ideas already discussed under this heading are reversible in so far as it does not matter whether the boss is stationary or rotating. Figs 146, 147, 148, 149 and 150 show suitable methods of locating a rotating shaft in a fixed boss.

The suggestions and examples given under this heading by no means comprise a complete list of methods of boss and shaft fixings and locations. The designer is at liberty to devise some means of his own or to combine some of the ideas shown. There are, however, several points still to be discussed regarding the detail design of bosses, bushes and shafts:

(*a*) all bosses should be chamfered on the bore to facilitate loading of the shaft or bush;

(*b*) all bushes should be chamfered on the outside diameter to facilitate insertion into the boss;

(*c*) all bushes should be chamfered on the bore to allow shafts to be loaded easily;

(*d*) all steps in shafts should have a radius in the corners. This avoids weakening the shaft. The chamfers on the bushes and bosses should be large enough to accept the radii on the shafts;

(*e*) radii should be provided in the corners of flanged bushes;

(*f*) when the shaft projects through the bush, the end of the shaft should be chamfered;

(*g*) chamfers should be provided on both sides of collars.

Fig. 152 shows these details on a suitable assembly.

176

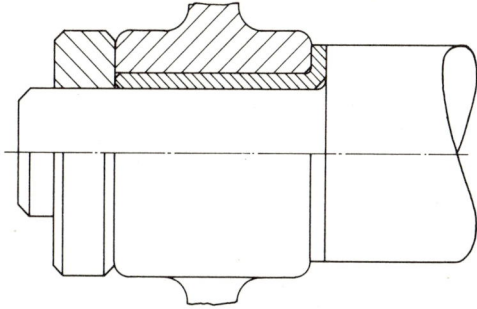

Fig. 152 Chamfers and radii in a shaft/boss assembly

Guides and slides

One essential requirement of a machine tool is a range of table or tool movements taking place along their mutually perpendicular axes. In a centre lathe the tool movements are controlled by the saddle, cross slide and compound slide. The movement is obtained by moving a sliding member along a guideway of some description. In machine tools the sliding members and guides are in the form of vees.

Other mechanisms often require slides and guides in some form or another. Feed screws, pneumatic or hydraulic cylinders and lever mechanisms provide movement to a sliding member which moves along a slideway or guideway.

In its simplest form, the slideway or guide can be a circular or square bar fixed firmly in supports. The sliding member should be sufficiently long to obtain a satisfactory sliding action. A length equal to 3–6 × bar diameter or side length is considered adequate.

Fig. 153A shows a single round bar with a sliding member. To

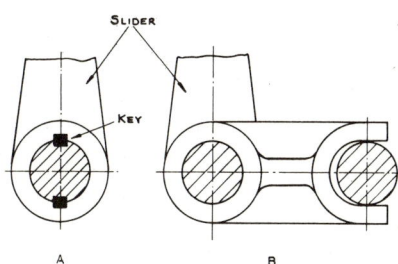

Fig. 153 Circular bar guides

prevent the sliding member changing its angular position two keys are fastened to the shaft. The angular position can be maintained by using a splined shaft and matching splined bore on the slider.

Fig. I53B shows how the angular position of the slider can be kept by using two plain circular bars. The slider has an arm fitted to it and is slotted to allow for misalignment of the guidebars without affecting their useful service.

The bores of the sliders may be fitted with bushes of cast iron or phosphor-bronze which can be replaced when worn.

The square bar guide can be used on many applications and it requires no machining or additions, to maintain the slider's angular position.

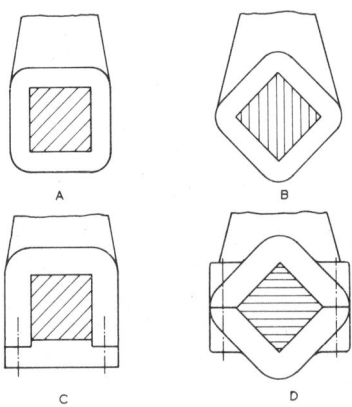

Fig. 154 Square bar guides

Fig. 154A shows the square bar in use as a guide.

Fig. 154B shows the bar turned through 45°.

In each case there is no provision for wear.

In Fig. 154C the cap, shown fitted, will eliminate wear in one direction.

In Fig. 154D the capping can eliminate wear in most cases.

Two circular or.square section bars can be used when a heavier duty slide system is required. The arrangements of the twin bars are shown in Fig. 155; the arrangements at A, B and C require accurate machining and alignment, whilst the arrangement shown at D requires only accurate alignment of the square bar at 45°.

There is no reason why bars of any section should not be used as guides.

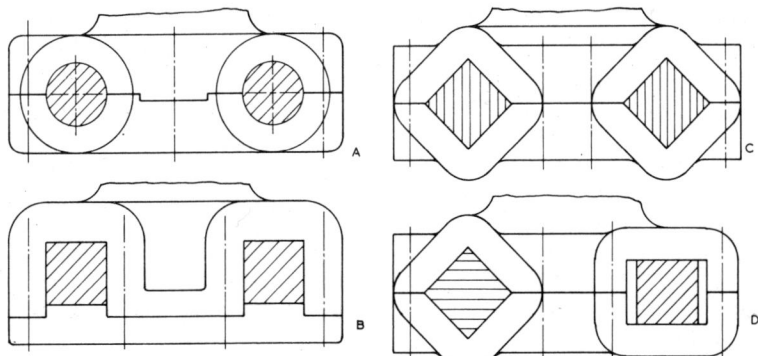

Fig. 155 Twin bar guides

One of the most popular parallel guiding units is the vee slide, the principle of which is shown in Fig. 156. It consists of two sliding vee pieces, a gib strip and some means of adjusting the strip. The arrangement shown is suitable only for light duty; the gib strip is prevented from moving by the peg ended adjusting screws. A more robust arrangement is shown in Fig. 157; after adjustment of the gib by the side screws, the gib is located by the vertical screws. The angle P shown, is usually 55° but angles of 60° and 45° are also used.

Fig. 158 shows an adjusting strip the back face of which is inclined. As the locking screws are tightened the strip is drawn up the taper and forced sideways to take up any play; the screw holes should be slotted to give freedom to the screws.

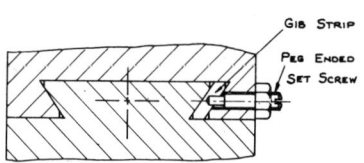

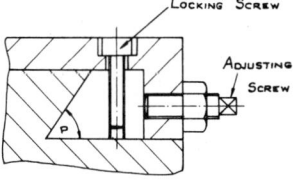

Fig. 156 Vee slide arrangement   *Fig. 157 Vee slide arrangement*

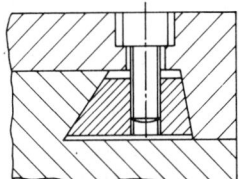

Fig. 158 Alternative gib strip arrangement

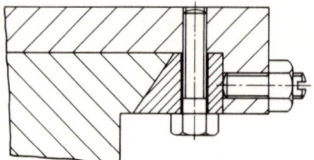

Fig. 159 Overhung vee slide

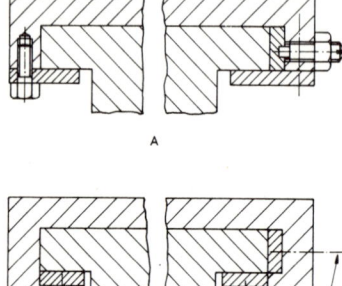

A

B

PEG ENDED
SET SCREWS

Fig. 160 Square slides

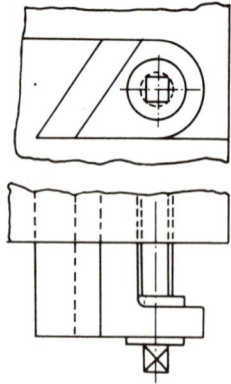

Fig. 161 Use of a tapered strip

Another alternative is shown in Fig. 159, and finds use on overhung slide applications.

Square slides are often used in overhung applications. They are more robust than the vee slides and can be used for correspondingly heavier duty. Fig. 160A shows a square slide arrangement; a gib strip is adjusted by peg ended screws and the moving slide is prevented from lifting by means of the two cover plates.

Fig. 160B shows a further development where adjustment is obtainable in both directions. Peg ended screws prevent the strips from moving.

The parallel strip may be replaced by a tapered strip, which can be used on square and vee slide applications. Fig. 161 shows a tapered strip fitted to a vee slide; adjustment is made by means of the screw. The right hand side of the strip mates with a taper of 1:50 on the moving part of the assembly. When the strip is fed in, the play is eliminated by the wedging action of the tapers.

Fig. 162 shows a typical assembly of a vee slide actuated by a screw: this arrangement is often used on machine tools.

Power transmission

In all forms of machinery, motion is supplied by a prime mover and transmitted by shafts which are connected together directly or

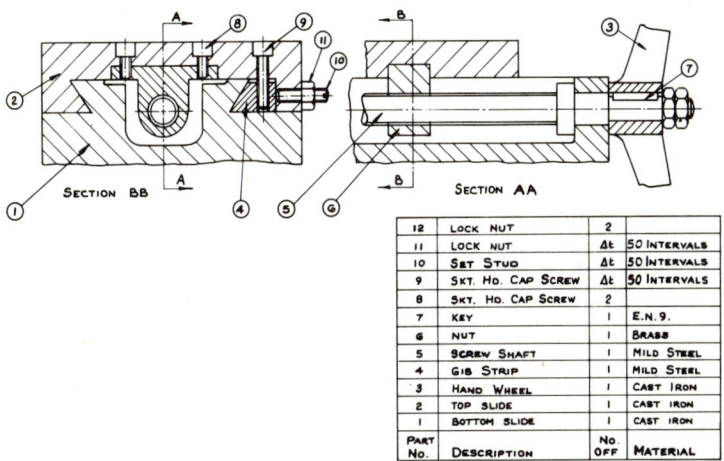

Part No.	Description	No. Off	Material
12	Lock Nut	2	
11	Lock Nut	At	50 Intervals
10	Set Stud	At	50 Intervals
9	Skt. Ho. Cap Screw	At	50 Intervals
8	Skt. Ho. Cap Screw	2	
7	Key	1	E.N.9.
6	Nut	1	Brass
5	Screw Shaft	1	Mild Steel
4	Gib Strip	1	Mild Steel
3	Hand Wheel	1	Cast Iron
2	Top Slide	1	Cast Iron
1	Bottom Slide	1	Cast Iron

Fig. 162 Assembly of a vee slide

across spaces. Some efficient means of transmitting the supplied power is required and devices enabling the drive to be disconnected at will or the speed to be altered, are required. Many such devices are available commercially and manufacturers catalogues should be consulted to establish the size of unit required to transmit a given power. A brief survey of power transmission units follows but the list is by no means complete.

Couplings

Couplings are used to connect two shafts directly to each other. Such instances occur when an electric motor is connected to a gear box input shaft. The shafts need to be separate for assembly and maintenance purposes, but require connecting to affect a useful unit. The gear box output shaft must also be connected to some other shaft on the machine it is to drive and again, some form of coupling is required.

Couplings fall into three groups, namely: rigid, flexible and universal.

Fig. 163 shows a rigid flange coupling which has been standardized (B.S. 2715:1956). It is admirably suited to power transmission when the two shafts which it is to connect are in alignment. The two halves are made from cast iron and are keyed with tapered keys to the shafts. A number of fitted bolts which hold the two

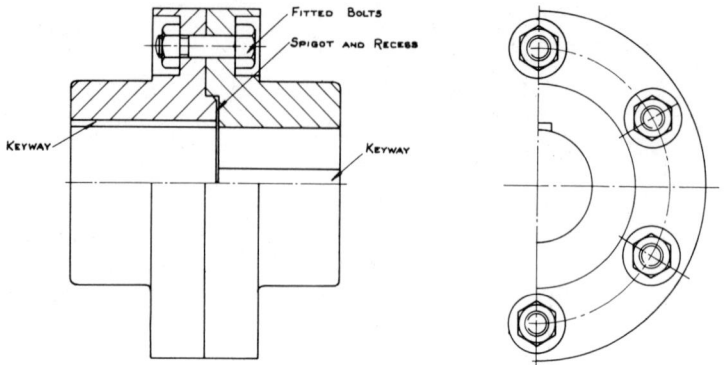

Fig. 163 Rigid flange coupling

flanges together, transmit power from one coupling half to the other. The two halves of the coupling are kept in alignment by means of a spigot on one half, which fits into a recess on the other half. Care must be taken to ensure that the shearing stress on the bolts caused by the transmitted torque is taken by the shank of the fitted bolt and not by the threaded portion.

When the alignment of the shafts cannot be ensured, some form of flexible coupling should be used. There are many forms of flexible coupling available commercially and manufacturers' catalogues will show the variety and range obtainable. Most flexible couplings will cope with small errors and variations in alignment and will damp out vibrations or starting shock which may occur in an assembly. Fig. 164 shows a pin type flexible coupling in which

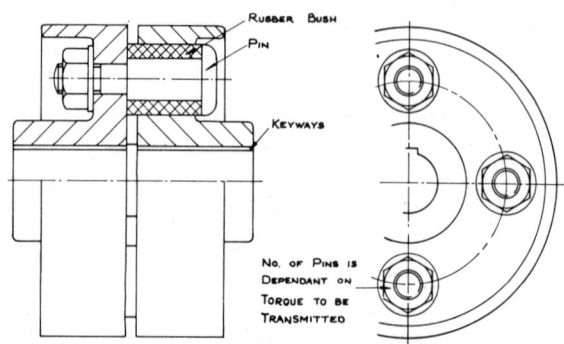

Fig. 164 Pin type flexible coupling

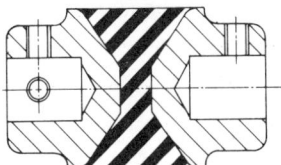 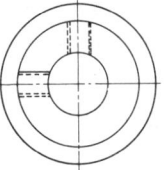

Fig. 165 Bonded rubber flexible coupling

the special bolts carry rubber bushes as shown. The rubber bushes engage in holes in the other coupling half and affect a drive.

An effective flexible coupling is shown in Fig. 165; the rubber is bonded to the metal coupling halves, which are held to the shafts by grub screws or keys. This type of coupling is capable of transmitting up to 12 kW and has the normal advantages of a flexible coupling plus a reduced overall diameter.

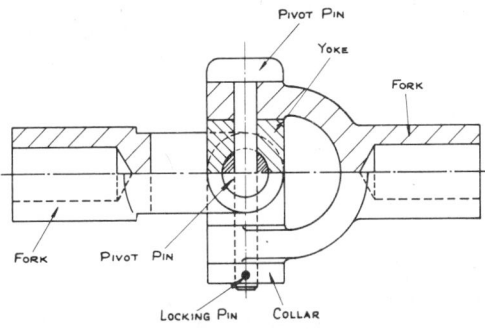

Fig. 166 Universal joint

When the angular misalignment of the two shafts exceeds two or three degrees, the flexible coupling is no longer suitable and a universal coupling or joint should be used, as shown in Fig. 166. It consists of two forked pieces which are connected to a yoke in the form of a cross. In smaller universal joints the yoke is replaced by a ball which has been flattened at the four points where it contacts the forks and the forks themselves are also much simpler in construction. This type of universal joint is shown in Fig. 167. The angle between the two shafts can be varied even while the shafts are in motion. When a universal joint is used, the angular velocity of the second shaft is not consistent. However, if two joints are used in a layout similar to that shown at Fig. 168A, where the angles marked K are equal and the joints on the middle shaft are

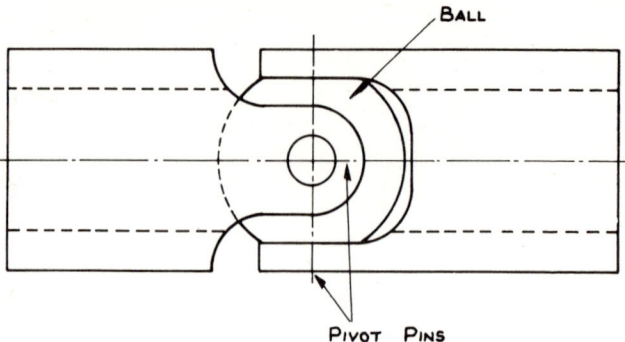

Fig. 167 Ball type universal joint

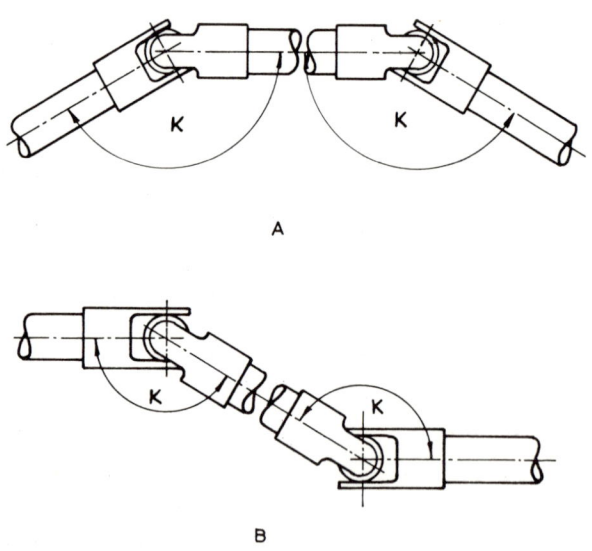

Fig. 168 Use of universal joints

in alignment, the angular velocity at the third shaft will be constant and identical to that of the first shaft. Fig. 168B shows another arrangement where the angles between the shafts are equal and the angular velocities of first and third shafts are constant. The system outlined in Fig. 168B is in use as the propeller shaft of motor vehicles connecting the gear box to the rear axle. It will be realized

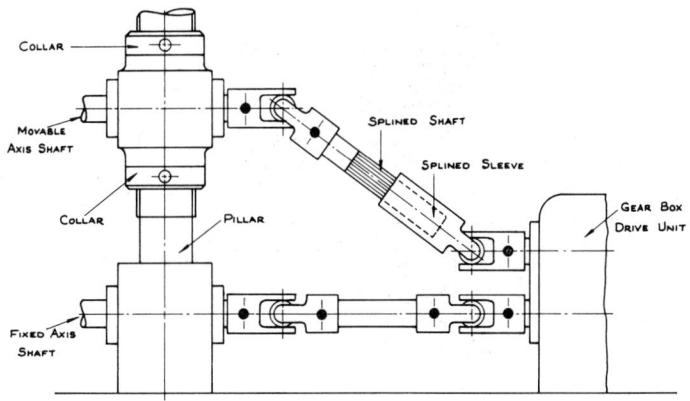

Fig. 169 Typical drive unit with universal joints

that the angles of the universal joints are constantly altering according to the load on the vehicle, road conditions and springing. In many industrial assemblies, this variation in angle of the universal joints results in variation of the shaft length and this is usually accommodated by means of a splined shaft and sleeve. Fig. 169 shows an outline of the drive to two shafts, which have a variable centre distance. The lower shaft is fixed on its axis and the upper shaft is lowered or raised by means of the screw collars on the pillar according to the adjustment required. The system of using a splined shaft and sleeve is often used as a coupling when the shafts can be located elsewhere or when the length of coupled shaft is variable.

Clutches

A clutch is a means of disconnecting and reconnecting a drive at will, most modern clutches transmit power by means of friction. When the revolving member of the clutch is gradually brought into contact with the stationary member, a drive is affected without shock. Such devices are in use on machine tool drives and motor cars.

Fig. 170 shows a section through a plate clutch (simplified). The shaft carries the housing B firmly keyed in position against a collar; friction plates C are prevented from rotating in the housing but are free to move axially; friction plates D are likewise prevented from rotating but are free to move axially on the fixed member E which is firmly fixed to shaft F. When an axial force is applied to

185

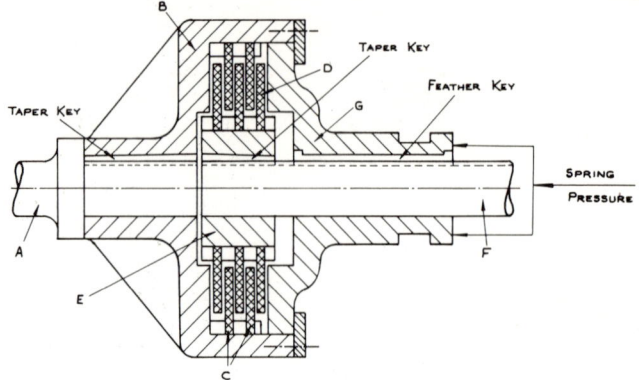

Fig. 170 Multi-plate clutch

the pad G, the friction plates are pushed together and can transmit power from shaft F to shaft A. The pad is provided with a groove into which a mechanism, Fig. 174, is fitted to enable it to be actuated. The housing, fixed member and pad are usually made from high grade cast iron whilst the plates can be made from cast iron or cast iron faced with a high friction material such as that used for brake linings.

Another form of clutch suitable for connecting shafts in axial alignment is shown in Fig. 171. This is known as a cone clutch and consists of two iron castings, one, A, is keyed firmly to its shaft whilst the other, B, is free to slide. B is actuated by the shifting mechanism shown in Fig. 174. A spring supplies sufficient force to keep the two halves of the clutch in close contact. The included angle of the cone can vary depending upon whether or not the male cone is covered with some high friction material. If left plain,

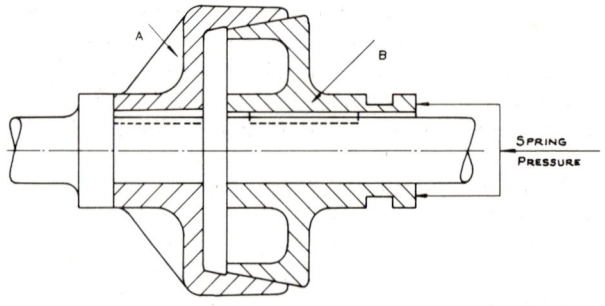

Fig. 171 Cone clutch

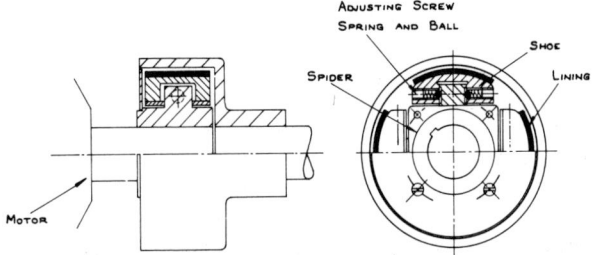

Fig. 172 Centrifugal clutch

the angle should be 20° but covered, the angle is increased to 36°.

Both plate and cone clutches require considerable axial force to keep the clutches in working engagement and the end thrust thus set up must be accommodated in the shaft bearings.

Fig. 172 shows a centrifugal clutch which is used to relieve the starting load on a motor or other prime mover. The motor can start under no load conditions, the working load being applied gradually and automatically when the motor has reached the required speed. The half of the clutch fastened to the prime mover's shaft is in the form of a spider which carries four shoes fitted with high friction material linings. As the speed increases the shoes move outwards under centrifugal force. When the shoes are free, the load is taken up gradually from starting and is suitable for general use but when motors would otherwise be started with a heavy load, the shoes are spring controlled. This ensures that the load is taken up at 75% of the motor's maximum speed.

Fig. 173 shows a device which is known by several names viz: claw clutch, claw coupling or dog clutch. The fixed half of the clutch is keyed firmly to its shaft whilst the other half is free to

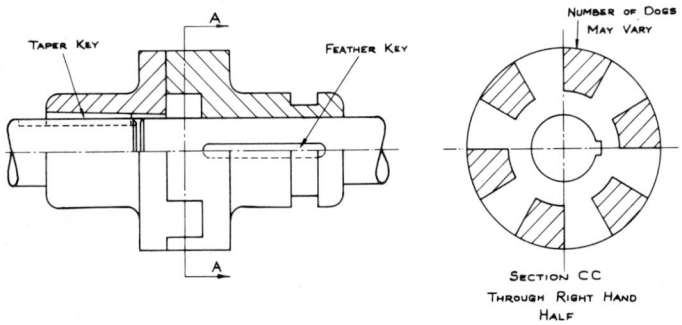

Fig. 173 Dog clutch

slide along its shaft; a feather key is used to prevent rotation. The sliding half is actuated by a shifting mechanism but the machinery must be stopped to engage the clutch. Disengagement can take place while the shaft is rotating. Note that one shaft projects into the other half of the clutch. This supports the shaft while it is out of engagement. A bush is often used to reduce friction but adequate lubrication is necessary in either case.

This type of clutch is sometimes used to connect the feed and lead screws of machine tools to their drives; in this case the engaging and disengaging is done by hand whilst the shafts are stationary.

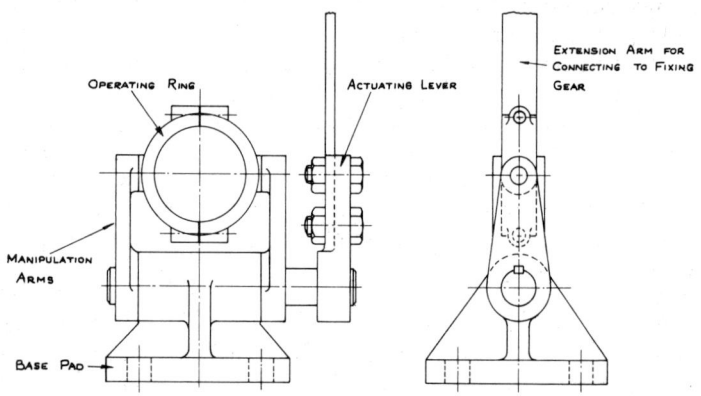

Fig. 174 Clutch shifting mechanism

Fig. 174 shows a shifting mechanism for clutches. The operating ring, which is split, fits into the groove provided in the clutch and is manipulated by two arms keyed firmly to a shaft supported in a base pad. The shaft is extended to carry an actuating lever. The extension arm is usually connected to some form of fixing arrangement to enable the clutch to be held out of engagement against spring pressure. This may consist of a ratchet arrangement or plunger and slot assembly.

Belt and Chain Drives

When power is to be transmitted between two shafts which are parallel to each other but are not in alignment, they can be connected by a belt or chain drive.

Endless belts of various dimensions having a cross section as shown in Fig. 175 are known as V-belts and are covered in B.S.

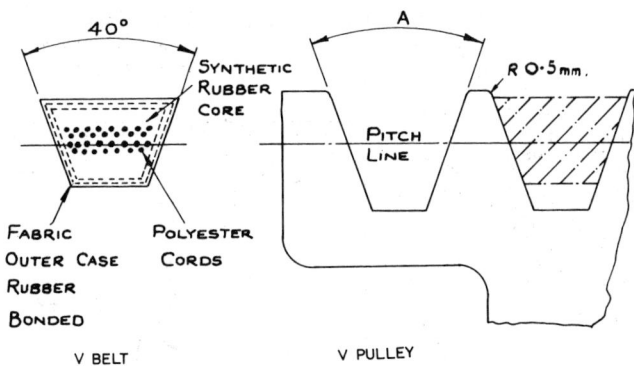

Fig. 175 Vee belt and pulley

1440: 1968. They can be used singly or in multiples and run in pulleys grooved to match the belts. Pulley diameters and belt lengths are calculated on a pitch line which lies at half the depth of the belt. The driven shaft can be operated at a different speed to the driving shaft by using pulleys of different diameters. The diameters of the pulleys are inversely proportional to the speeds at which the shafts are to rotate.

A form of belt drive which can be used when a positive, non-slipping drive is required is shown in Fig. 176. This is often known as a timing belt since once the angular relationships of the two shafts has been fixed, they will remain constant. The belt is basically flat, but has transverse moulded ribs at regular intervals which engage in pulleys having matching slots cut into their rims.

Chain drives are familiar to most people since they appear on bicycles and motor cycles. They have a large industrial use as a positive means of transmitting power. Chain is available as a single link chain known as 'simplex', but is often made in the form of a double link or treble link chain known as 'duplex' and 'triplex' accordingly. The chain wheels or sprockets are toothed to carry the chain and their speeds are inversely proportional to the numbers of teeth on the wheels or sprockets. Many different types of chain drive have been developed, such as inverted tooth chains for high speed drives, and long link and conveyor chains for conveyor, elevator and other types of mechanical handling equipment.

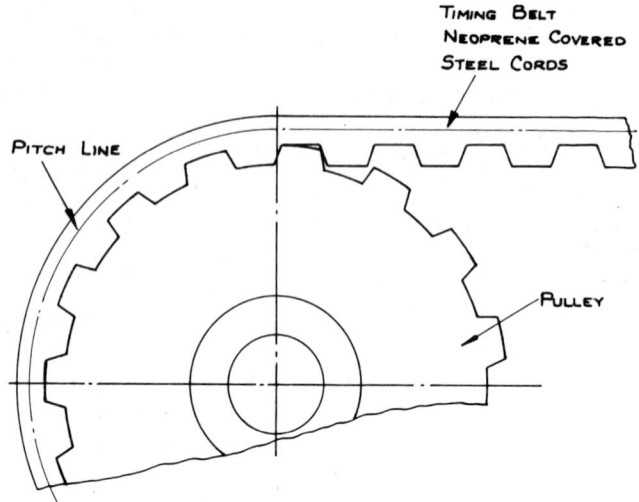

Fig. 176 Timing belt and pulley

When a belt or chain drive is required for a specific application, the manufacturers' catalogues should be consulted. Many firms carry stock drives and by quoting driven shaft speed, driving shaft speed, shaft centre distance and horsepower to be transmitted, a stock drive can be forwarded by return. These are often cheaper than special drives. It is good practice to consider a stock drive at the design stage rather than design and then try to fit a stock drive to the design.

Choice of constructional method

The methods of producing raw component parts for subsequent machining into finished components include casting, forging, fabrication by dowels and screws, fabrication by welding, press work and die casting. The type of product available by means of presswork and die casting is limited to thin sheet metal forms by press work and small light low melting point alloy castings by die casting. A large number of components can be produced by machining from solid bars of cast iron, mild steel or other materials. The choice of production method depends upon the number to be produced, the size, shape, complexity and purpose of the component, the machining capacity available, and the material from which the component is to be made. Often, after due consideration of the factors involved, the choice of production method is still open. This occurs with the vast majority of components and the designer is left to select from fabrication, casting and forging. Fabrication by means of screwing and dowelling pieces of material together is a very expensive production method, even though the finished article is accurately made; this method is best left to the production of jigs and fixtures. If a component is to be fabricated it will usually be done by welding. The range of materials which can be fabricated in this way include mild and medium carbon steels, brasses, copper, aluminium alloys and thermo-plastics. Sections of varying thickness can be welded together without difficulty, but care should be taken to ensure that the heat generated by the welding process does not warp the finished fabrication. If this is likely to occur then sufficient material should be allowed for machining off at a later stage. Where much welding has been done on a component, it is usual to provide a stress relieving heat treatment which is recommended as sound practice.

Fabrication is eminently suitable as a production method when stock size material can be used to build up a component. Tubes, plates, bars, rods and structural sections can all be utilized to fabricate a component. It often happens that a replacement part for a piece of equipment or machinery is needed. The original part might be a casting or forging but the replacement can be quickly fabricated from stock sections. Replacement brackets, levers, baseplates etc. can easily be produced by this method.

In recent years fabricated construction has found wider application. Strong, light frameworks are now being produced which would have previously been cast, and future years may see machine tool beds fabricated from steel. Already one machine tool manufacturer has had considerable success using fabricated structures. There are, however, many applications to which fabrication can naturally be applied. Pressure vessels, pipework, ducting and structural steelwork lend themselves to this method of production. When stainless steel is specified for a particular item, it is usually found to be fabricated by welding.

Although plastics are normally moulded to their required shape, recently developed techniques enable thermo-plastic materials to be welded. The advantages of cheapness and absence of dies when only a few items are required are obvious enough.

The casting of molten metal into a mould of pre-determined shape has been carried out since man first discovered metals.

Modern casting techniques enable sound, non-porous castings to be produced quickly and cheaply, and the advent of modern high duty and spheroidal graphite cast irons has widened the scope of iron casting into fields which they would not have previously entered. Modern non-ferrous castable alloys have also increased the applications of cast products.

Whilst the castings themselves are cheap to produce, the cost of the materials and the patterns must also be taken into consideration. The production of a wooden or metal pattern is costly and when it is finished it is only suitable for producing a casting similar in shape to the pattern. If only one casting is required, the cost of the pattern must be an additional expense but it can, of course, be spread if more than one casting is required. Generally speaking casting should be employed for the production of batches of components. There are, however, occasions when this cannot be applied, for example if a special purpose machine is manufactured then a single casting only might be required.

The use of cast iron as an engineering material need not be limited to special castings. Indeed, several firms will supply solid or hollow cast sticks of iron up to approximately 225 mm diameter. These can be used to produce small cast iron items by machining.

When specifying that an item should be produced by casting, care should be taken to ensure that it is possible for the component to be cast. A quick survey of any available casting will apparently reveal that the foundry master is a miracle worker, but this is not so. If at all possible, the foundry should be consulted during the design stage of any proposed casting, to ensure a satisfactory result.

There are many components which are easier to cast than to fabricate, e.g. motor vehicle gear boxes, differential gear boxes and engine blocks; these would be very difficult to produce as fabrications without radically altering the shape.

The production of components which have three-dimensional curved shapes is eminently suited to the casting technique, as success depends on the pattern. To produce a fabrication having similar curved features, one has to resort to hot pressing or hot spinning processes, which call for specially shaped dies and tools and are limited to repetitive production of similar components. However, if the component has two-dimensional curved features such as simple bends and radii, the fabrication technique can be utilized as this type of work can be done on standard machinery.

Chills are often used in the moulds of iron castings to produce hard surfaces in the component and obviously this cannot be done on a steel fabrication. There is a welding technique however, when a pad of stellite, carbide, stainless steels can be quickly and easily deposited in any desired position.

Forgings from a few grammes up to 200 tonnes can be produced by use of hand, eye and some relatively simple tool for supplying force, usually a mechanical hammer or hydraulic press. The work varies from simple pipe clips made at a blacksmith's forge to a large crankshaft made after several hours work by a gang of men under the leadership of a forgemaster.

On the other hand there is the forging technique known as *drop-forging*, used for the quantity production of relatively small items such as hammer heads, spanners etc. This technique is a cheap way of producing forged articles providing the cost of the dies can be spread over large numbers of components.

The hand techniques are ideal for producing one or two components provided that their shapes are not too complex. Forgings are well known for having a homogeneous grain structure not found in components produced by other methods; this gives a forged component superior strength in applications where severe stresses are likely to occur.

Hand forging techniques require great skill and take considerable time; thus a component which is to be forged is likely to be costly. Forging should therefore be limited to components which will benefit from the use of this technique. Many items which were forged in the past are now cast in high duty and malleable irons, or fabricated, as these are cheaper methods.

Many components will suggest their own production method to the designer, but inevitably there will be occasions when the choice is not so obvious or when apparently either of two methods would give an adequate solution. When this happens, the choice must be left to the designer after he has considered the cost of each technique, the available time in which the component is to be made, and the number of components required.

Conclusions

It is hoped that the problems which follow will promote much class discussion and individual research as the writers feel that these are far better methods of studying the constructional and operational features of machinery and equipment, than the laborious methods of note taking and sketch making.

Engineering Construction and Materials

In all the exercises, frequent reference should be made to B.S. 308: 'Engineering Drawing Practice'. Tolerances when asked for should be extracted from B.S. Data Sheet No. 1 'Limits and Fits': 'Primary Selection of Fits'.

Worked example

Fig. 177 shows the shaft output of a variable speed gear box and the lower end of a link which must be connected in such a way

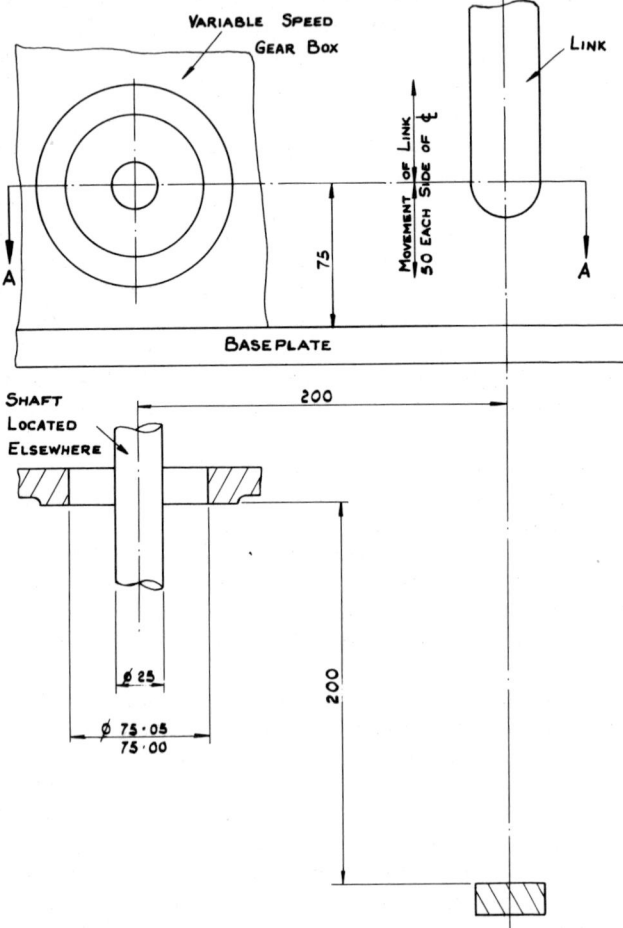

Fig. 177 Problem for worked example

that the rotary movement of the gear box shaft is translated into a straight line reciprocating motion at the upper end of the link. The gear box output shaft revolves at 150–250 rev/min and the link is to reciprocate at 30–50 times/min.

The problem is to devise a suitable drive arrangement to operate under the conditions stated. From the information given on Fig. 177 one can presume that the hole size in the gear box is fixed but the shaft can be altered to suit any fittings and bearings that may be required. The problem requires that the upper end of the link is to move in a straight line and, from this, one may deduce that the lower end of the link can move at random as long as the upper end moves in a straight line.

When faced with such a problem as this, the designer must evolve a scheme which will fulfil the requirements, be reasonably cheap to make, be easily maintained and operate safely.

Fig. 178 shows the writers' solution to this problem. In order to obtain the desired link motion, the gear box is made to drive a crank which is connected to the link. One revolution of the crank will cause the upper end of the link to reciprocate once. The crank is made to rotate by driving the shaft, to which it is connected, from the gear box. The gear box shaft revolves at 150–250 rev/min, and the link reciprocates at 30–50 times/min, hence the crank revolves at 30–50 rev/min, and a drive having a reduction of 5:1 is required. A chain drive has been chosen as this will give an accurate reduction drive which is positive in action. The speed is relatively slow and if the loading is not severe, the simplex chain would be adequate. Gears or vee belts could have been used for this drive but were not selected because the gears would be more expensive than chain and the vee belts may have had a tendency to slip.

Adjustment of the chain drive is affected by means of a single jockey pinion which lifts the chain to give a greater angle of lap on the driving pinion. The jockey pinion is bushed and is free to rotate on its shaft, which can be raised or lowered in the bracket by slackening the holding nut. That part of the shaft which passes through the bracket is flatted so that the shaft does not rotate when the nut is turned. The jockey bracket is fabricated by welding from mild steel plate and the seating faces are machined square to each other when welding is completed. The offset web provides the bracket with sufficient stiffness. In order to align the jockey with the chain the brackets' fastening holes are provided with slots, through which pass studs from the base-plate.

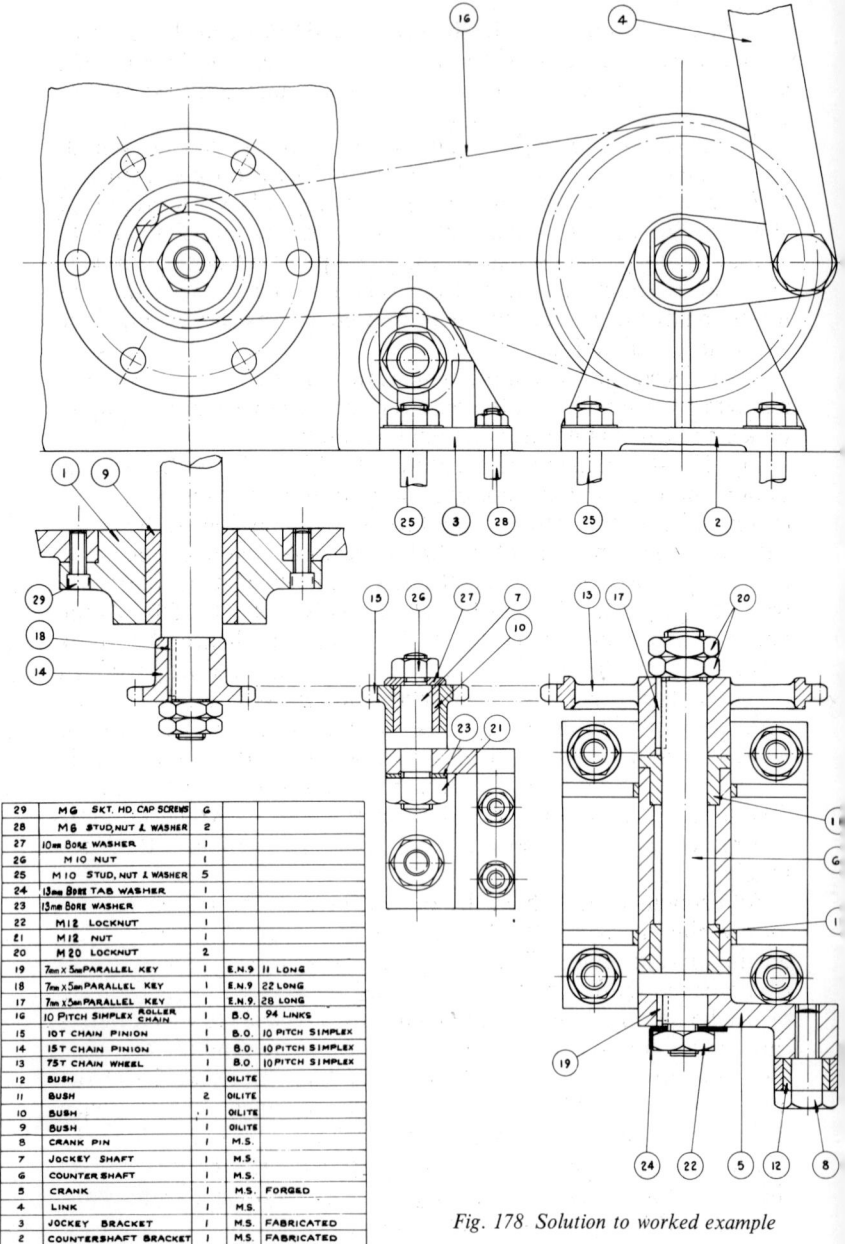

Fig. 178 Solution to worked example

29	M 6 SKT. HD. CAP SCREWS	6		
28	M 6 STUD, NUT & WASHER	2		
27	10mm BORE WASHER	1		
26	M 10 NUT	1		
25	M 10 STUD, NUT & WASHER	5		
24	13mm BORE TAB WASHER	1		
23	13mm BORE WASHER	1		
22	M 12 LOCKNUT	1		
21	M 12 NUT	1		
20	M 20 LOCKNUT	2		
19	7mm X 5mm PARALLEL KEY	1	E.N.9	11 LONG
18	7mm X 5mm PARALLEL KEY	1	E.N.9	22 LONG
17	7mm X 5mm PARALLEL KEY	1	E.N.9	28 LONG
16	10 PITCH SIMPLEX ROLLER CHAIN	1	B.O.	94 LINKS
15	10T CHAIN PINION	1	B.O.	10 PITCH SIMPLEX
14	15T CHAIN PINION	1	B.O.	10 PITCH SIMPLEX
13	75T CHAIN WHEEL	1	B.O.	10 PITCH SIMPLEX
12	BUSH	1	OILITE	
11	BUSH	2	OILITE	
10	BUSH	1	OILITE	
9	BUSH	1	OILITE	
8	CRANK PIN	1	M.S.	
7	JOCKEY SHAFT	1	M.S.	
6	COUNTER SHAFT	1	M.S.	
5	CRANK	1	M.S.	FORGED
4	LINK	1	M.S.	
3	JOCKEY BRACKET	1	M.S.	FABRICATED
2	COUNTERSHAFT BRACKET	1	M.S.	FABRICATED
1	BEARING HOUSING	1	C.I.	
PART Nº	DESCRIPTION	Nº OFF	MATᴸ	REMARKS

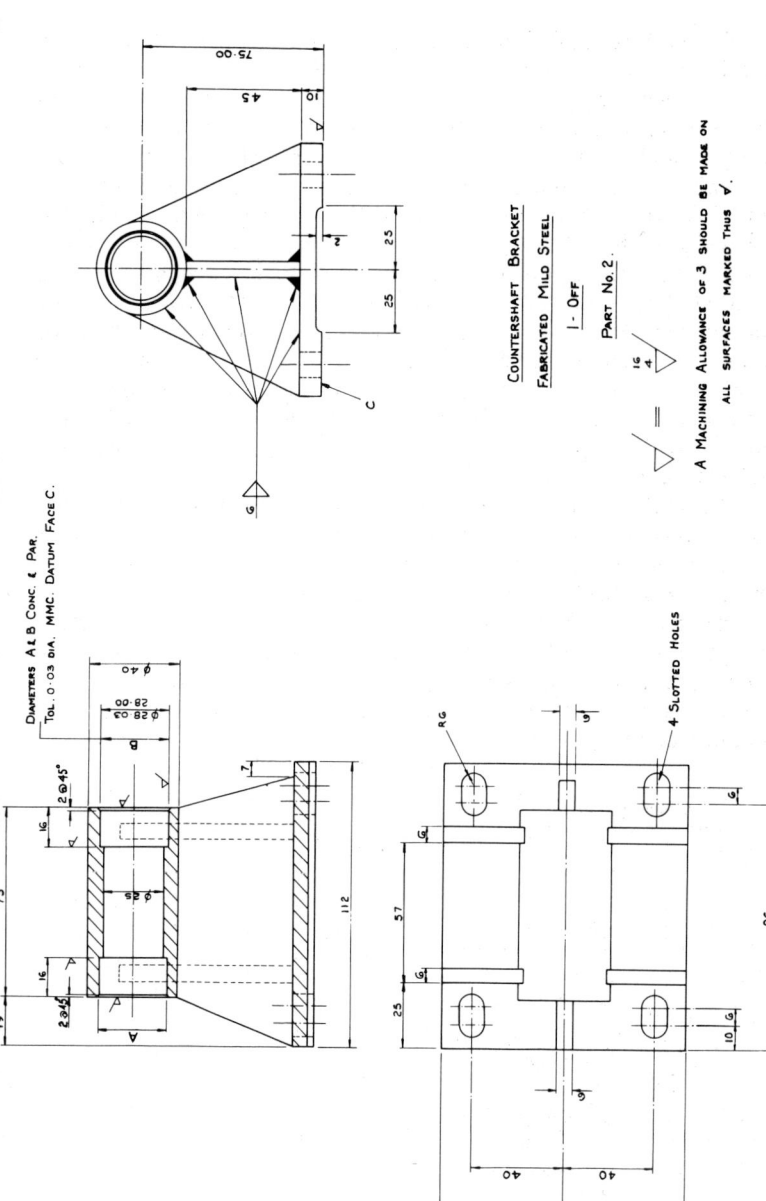

Fig. 179 Detail of countershaft bracket

The countershaft bracket is also fabricated by welding mild steel, plate and seamless tube of suitable wall thickness and a detailed working drawing of this item is shown in Fig. 179. This bracket is also provided with slotted holes for alignment purposes, and is held to the base-plate in a similar manner to the jockey bracket. The counter shaft bracket is provided with flanged bushes in which the countershaft revolves. At the crank end, the shaft is provided with an integral collar for locating purposes and this also locates the crank, which is held in place with a locknut and tab washer. The crank is forged from mild steel as this provides a more homogeneous component than if it were fabricated. A threaded crank pin secures the link, which is bushed, to the crank. At the other end of the countershaft, the chain wheel is fastened by means of a key and locknuts. This arrangement locates the shaft axially and provides a positive drive, together with adequate adjustment of end play.

The shaft from the gear box passes through a bush which has been pressed into a cast iron bearing housing held and located in the base of the gear box. The housing is detailed in Fig. 180 and the bush in Fig. 181.

The gear box shaft has been reduced in diameter to enable the chain pinion to be fitted and secured. In order to reduce maintenance to a minimum, all the bushes are made from sintered bronze which has been impregnated with oil. This material is available commercially as 'oilite'.

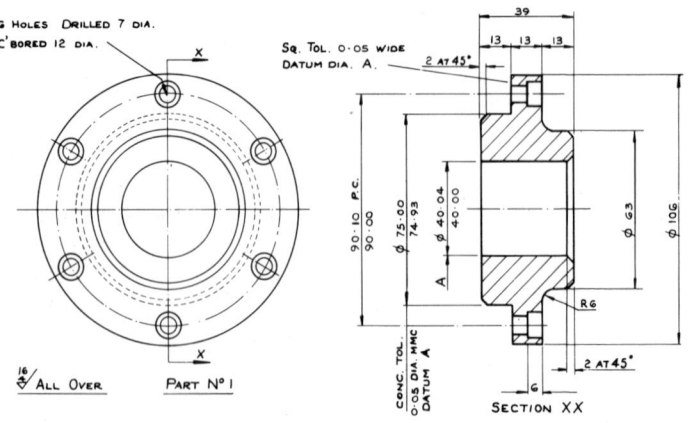

Fig. 180 Detail of housing

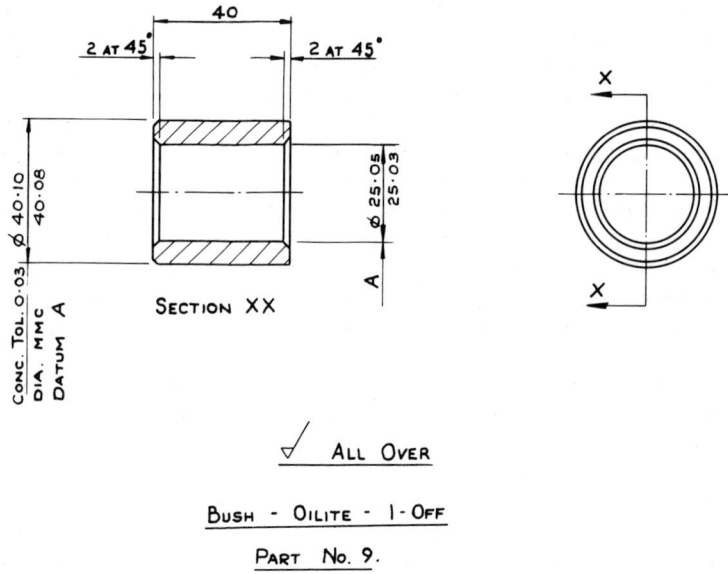

ALL OVER

BUSH - OILITE - 1 - OFF

PART No. 9.

Fig. 181 Detail of bush

Fig. 178 is a typical assembly drawing and shows little other than the component parts correctly fitted together. The parts list is an important item on this drawing as it names all the items required, states the material from which they are to be manufactured, and indicates which are standard items and which are to be purchased from an outside supplier. These latter are often known as bought out items and are indicated on the parts list as B.O.

Each component on the parts list is given a part number and this number is shown alongside the respective component on the drawing by means of an arrow, the number being placed in a circle; this is known as *ballooning*. This reference system enables the draughtsman to check that all the working drawings have been completed, that the materials and bought out items have been ordered and that work has been put in hand for the manufacture of the components. The parts list is also useful to the fitter who will eventually assemble all the components into a working unit as he can check that all the parts are to hand before commencing assembly. In addition, the assembly drawing gives him guidance as to the best assembly method and the manner in which the component parts are fitted together. The assembly drawing, however, is of no use to the people who are to manufacture the component parts.

No dimensions, tolerances, manufacturing methods or surface finish values are given and it is necessary to provide separate drawings, giving very detailed information, about the component, to the machinists, blacksmiths, fitters, welders and others who will manufacture the components. These detailed working drawings, known as *details* or *detail drawings*, should provide all the information likely to be required during the component's manufacture. Dimensions, of course, are extremely important and these should not cause confusion. Toleranced dimensions should be indicated clearly and care should be taken when selecting the limits of size so that the correct fits between mating parts are obtained. Reference to B.S. 4500A and B.S. 4500B is recommended. These are data sheets of limits and fits giving recommended limits of size for a variety of fits. The detail drawings should also give geometrical and positional tolerances of the components' features and should contain any notes necessary to enable manufacture to be completed. Surface finish and welding symbols should be added where necessary. It is usual to state on the detail drawing, the name of the component, the material from which it is to be made, the number of components required to be made and the part number.

Fig. 179 shows the detail drawing of Part No. 2, the countershaft bracket. Concentricity of diameters A and B with each other and parallelism with the base are indicated, together with surfaces to be machined, welding symbols and machining allowance.

Fig. 180 shows the detail drawing of the Part No. 1, the bearing housing. This component is machined all over and the locating spigot is shown concentric with the bore. The face which is to be clamped to the gear box side is shown square with the bore. Concentric and square tolerances are shown to achieve this.

Fig. 181 shows the detail drawing of Part No. 9, a bush. A geometrical tolerance is shown to limit the eccentricity of the outside and inside diameters.

In all three components, chamfers have been provided to assist fit and location.

On the detail drawings of Part Nos. 1 and 2, limits have been selected from the B.S. Data Sheet No. 1 'Primary Selection of Fits', B.S. 4500.

On Part No. 9, the limits quoted are those suggested by the manufacturer of the material from which the bush is to be made.

EXERCISES

1. Fig. 182 (A) shows the sketch detail of a proposed adjustable tool holder/tool post for a centre lathe. Fig. 182 (B) shows part of the lathe capacity chart with the dimensions which the tool post must meet.

   Make a neat assembly drawing showing the tool post/tool holder in its completed form, using your own judgement to decide the general dimensions and details.

   (C.G.L.I.)

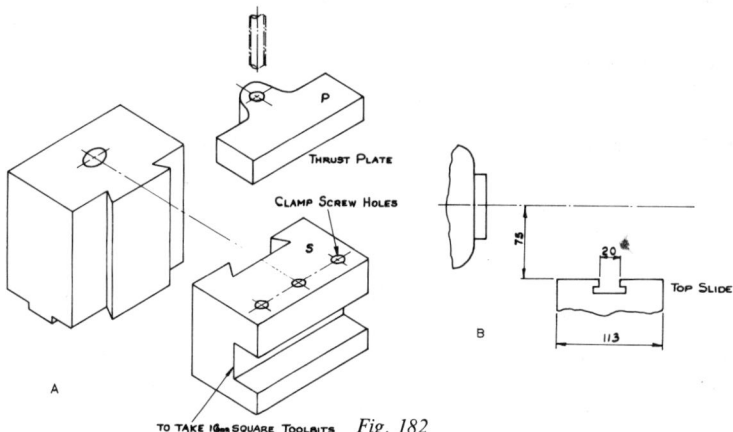

TO TAKE 16mm SQUARE TOOLBITS *Fig. 182*

2. Make a working drawing of the thrust plate P in Fig. 182, showing all the dimensions and instructions necessary for manufacture. Put limits on *two* dimensions only, and show *one* example of a geometrical or positional tolerance.

   (C.G.L.I.)

3. Fig. 183 shows some details of a gear box end plate, the driving shaft and a section of a pulley for a two speed drive by vee belt. The shaft is required to run at speeds of 200 rev/min and 300 rev/min, with a belt speed of 110 m/min.

   Draw, full size, a view of the assembly, partly in section, to show the bearing complete with bronze bush, the provision for lubrication, the attachment of the pulley to the shaft and the guard arrangement. Provision for taking thrust and end location of the shaft is made elsewhere. Main dimensions should be included but machining tolerances need not be shown.

   (C.G.L.I.)

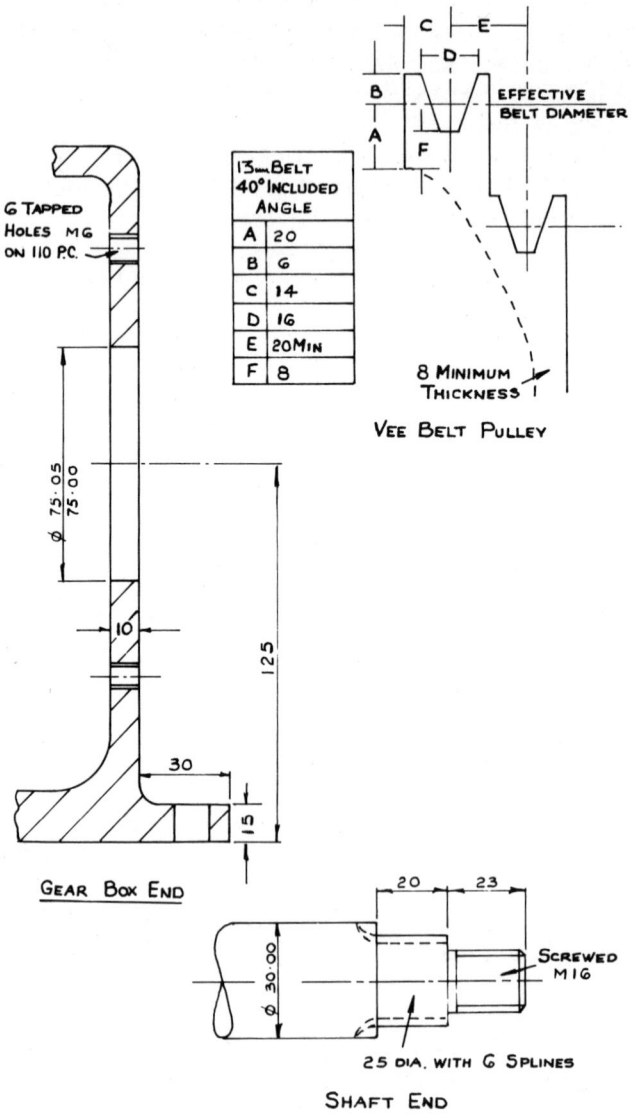

G TAPPED
HOLES M6
ON 110 P.C.

13mm BELT 40° INCLUDED ANGLE	
A	20
B	6
C	14
D	16
E	20 MIN
F	8

EFFECTIVE
BELT DIAMETER

8 MINIMUM
THICKNESS

VEE BELT PULLEY

Ø 75·05 / 75·00

10

125

30

15

GEAR BOX END

20   23

Ø 30·00

SCREWED
M16

25 DIA. WITH 6 SPLINES

SHAFT END

Fig. 183

4. From the arrangement drawing completed from Fig. 183, make detailed working drawings of three of the component parts, one of which should be the pulley. The detail drawings should include all dimensions and tolerances, notes and instructions necessary for the manufacture of the components.

5. Fig. 184 shows the main details of a small set of bench rolls designed for rolling soft metal strip. The bearing half at each end of the top roll can slide up and down in its slot to allow the roll to lift in order to vary the gap. The adjustment is obtained by means of two screws passing through the M22 tapped holes.

By means of sketches or drawings, suggest any arrangement of the two screws which would enable them to be turned simultaneously, or which would otherwise ensure that each end of the roll was adjusted equally in order to keep the gap parallel.

Note: it is not necessary to re-draw the whole assembly.

(C.G.L.I.)

6. The hand rolling mill shown in Fig. 184 has two rolls supported in half bearings. The top bearings can slide up and down in their slots to allow the gap to be varied. The end A of the bottom roll journal is extended to allow a handle to be fitted. The other end of each roll carries a spur gear. These mesh together so that the top roll is driven.

Draw two views of the assembly in correct relationship:
   (i) a cross section on the centre line of the rolls, looking in the direction of arrow 'X';
   (ii) an end elevation on the geared end.
The views should show the size and position of the gears, a suitable method of attaching the handle, and any means of lubricating the top bearing. It is not necessary to show any adjusting screws, the drawing may be to any suitable scale, and you may choose any suitable dimensions where none are given.

(C.G.L.I.)

7. Fig. 185 shows the general form of the bottom half of a casing for a worm and worm-wheel drive, with relevant details of the worm, worm-wheel and the bearings to be used. The wheel is mounted on phosphor-bronze bushes and the worm on ball bearings which are capable of taking all radial and thrust loads applied.

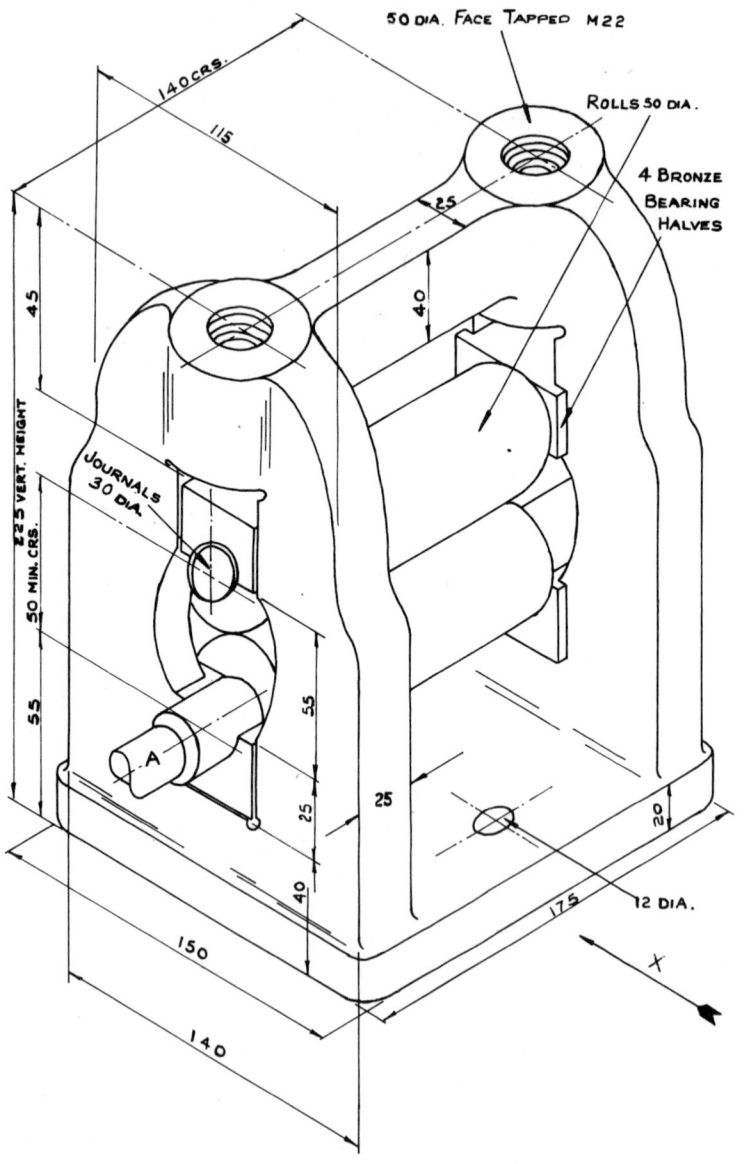

Fig. 184

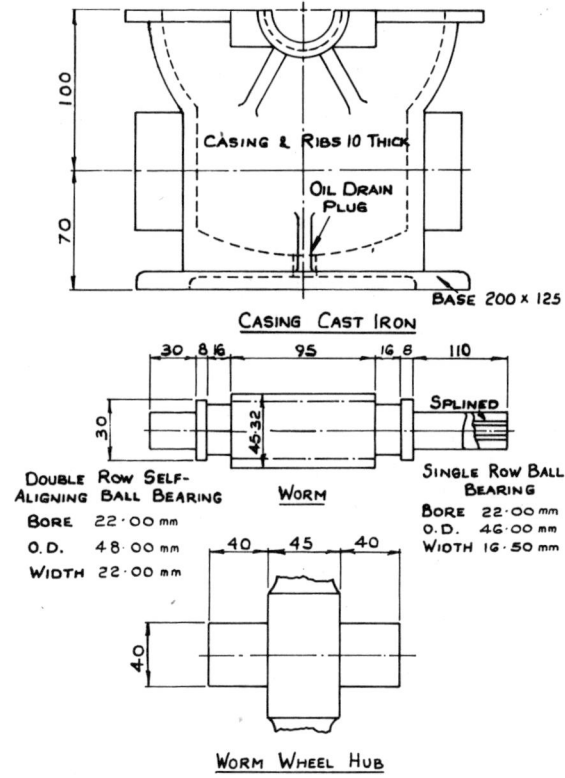

Fig. 185

Draw, full size, two views, part section as necessary, to show the general arrangement of the casing and its bearings including provision for transmitting the thrust load to the casing and for oil-sealing.

The worm and worm-wheel need not be shown and any dimensions needed, but not given, must be supplied.

(C.G.L.I.)

8. When the arrangement drawing of the gear box outlined in Fig. 185 has been completed, make detail working drawings of all the components with the exception of the gear box, worm and worm-wheel. All dimension, tolerances and other relevant information concerning the manufacture of these components should be shown on the drawing.

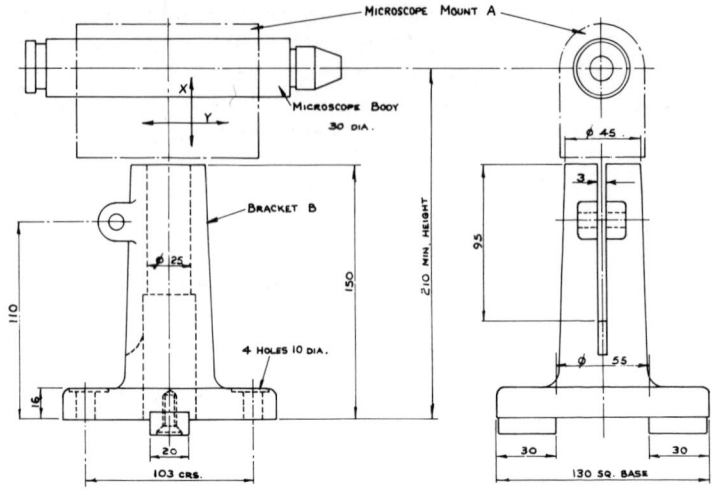

Fig. 186

9. A small microscope unit is required in order to study the teeth of cutters whilst being sharpened on a grinding machine. The conditions outlined in Fig. 186 show what is required. The bracket B is made to mount vertically on the grinder table. The microscope is a cylindrical unit which must be held very securely in a cast iron mount at A. The mount should be about 75 mm long. The microscope must have the following movements:

   (i) vertical rise and fall of about 50 mm. X direction;

   (ii) swivelling about the vertical axis;

   (iii) horizontal movement, direction Y, of about $\pm$ 10 mm from the centre line.

The horizontal movement should be operated by a rack and pinion or a screw and a suitable slide to ensure smooth action in a straight line. The vertical and swivelling movements should have provision for clamping.

   Show by drawing or sketching in good proportion, using suitable views and details, a suitable construction for the unit. Make a parts list, indicating suitable materials.

(C.G.L.I.)

10. The bracket B in Fig. 186 is to be supplied as good quality iron casting ready for machining. Make a working drawing showing all views, dimensions and instructions necessary for

manufacture. Suitable machining symbols and tolerances must be given, and use should be made of B.S. Data Sheet No. 1, 'Primary Selection of Fits'. Assume that the bore of the bracket and its shaft will have a sliding fit. Enclose the drawing in a standard frame similar to those recommended by B.S. 308: 'Engineering Drawing Practice'.

(C.G.L.I.)

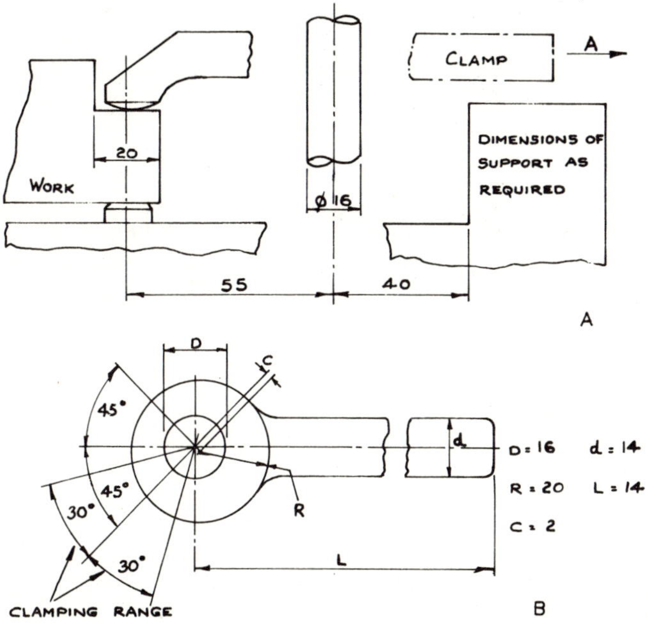

Fig. 187

11. The general plan of a certain fixture has been decided leaving certain details to be finished. One of these is a clamp, the outline conditions of which are shown in Fig. 187A. The clamp must slide in the direction A to release the work but must not rotate. It can be operated by a cam lock and the constants for a cam handle are given in Fig. 187B. Draw, full size, a side elevation of the completed clamping device, assuming any desired form or proportion for the various parts.

Balloon the parts and draw up a parts list stipulating the materials required and any necessary heat treatment.

(C.G.L.I.)

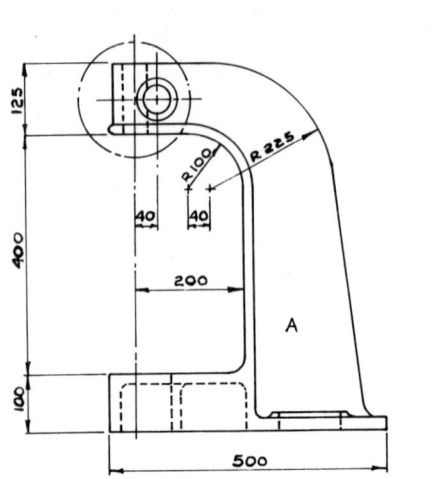

Fig. 188

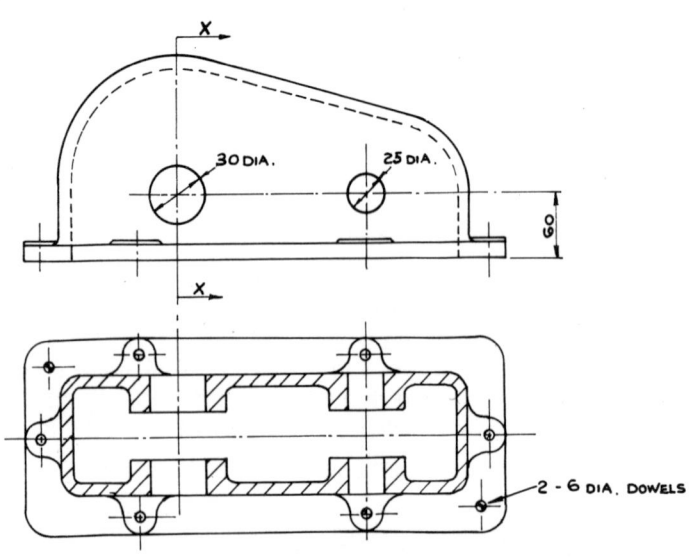

Fig. 189

12. Fig. 188 shows the frame of a mandrel press. Other main components to complete the assembly consist of

    (i) rack;

    (ii) pinion shaft;

    (iii) pinion shaft bush;

    (iv) lever arm (approximately 500 mm long).

Draw the elevation A with the remainder of the parts assembled to a scale of $\frac{1}{4}$ full size. Insert 6 principal dimensions.

Alongside this view draw, full size, an auxiliary view of the part of frame shown in the large circle. This view should be sectioned on XX, and with other parts assembled.

Gear teeth should be shown by standard convention and any dimensions not given should be estimated.

(C.G.L.I.)

13. Fig. 189 shows a gear casing for a small gear take-off drive. The gears are fixed to plain shafts 25 mm and 20 mm diameters respectively. The small gear has 39 teeth 2 mm module and the ratio is 3:5.

Draw, full size, the following views of the assembly:

    (i) elevation—outside view;

    (ii) end elevation in section of XX.

The views should show clearly the method of fastening the gears to their shafts, provision for lubrication and the method of preventing end float of the shafts. A toleranced centre distance of the shafts should be clearly indicated and a parts and materials list should complete the drawing.

(C.G.L.I.)

14. With reference to the completed assembly for Fig. 189, make detailed working drawings of the shafts, bushes and gears. The drawings should show all dimensions, tolerances, machining marks and instructions necessary for manufacture.

15. Fig. 190 shows three components of a cone clutch drive for a small lathe. The shaft is carried by a ball bearing, located in a bearing housing fastened to the gear-box casting by 6 Allen screws. A pulley assembly mounted on the shaft has the mating internal cone formed within it and the size and position of this pulley is shown in chain-dot line. A thrust rod pushes the driven cone into the driving cone and several minor components complete the main assembly.

Draw the completed assembly which allows the pulley to

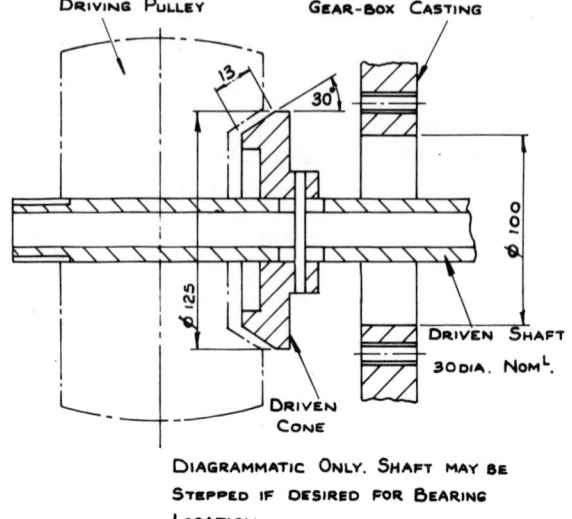

Fig. 190

rotate on the driven shaft and the driven shaft to rotate in the housing secured to the gear-box casting.

Note: the operation of the push rod need not be shown.

(C.G.L.I.)

16. Fig. 191A shows the main details of the body of a small gear pump for oil or suds. The driving gear has the form shown in Fig. 191B and is integral with the shaft, which carries a chain sprocket when operating. The gear shaft must pass through a suitable gland to prevent leakage. The pump is closed by a simple cover, the inside face of which must have a clearance not exceeding 0·05 mm between it and the gear faces in order to ensure satisfactory operation.

Draw a general arrangement of the pump as it would appear when fully assembled showing the gland, cover, gears in position and a sprocket mounted on the shaft.

Dimensioning is not required, but the various parts should be balloned and cross referenced with a parts list.

(C.G.L.I.)

17. Fig. 192 shows the body and quill of a spring loaded tailstock. A lever operates the quill and the lever ratio is 3:1.

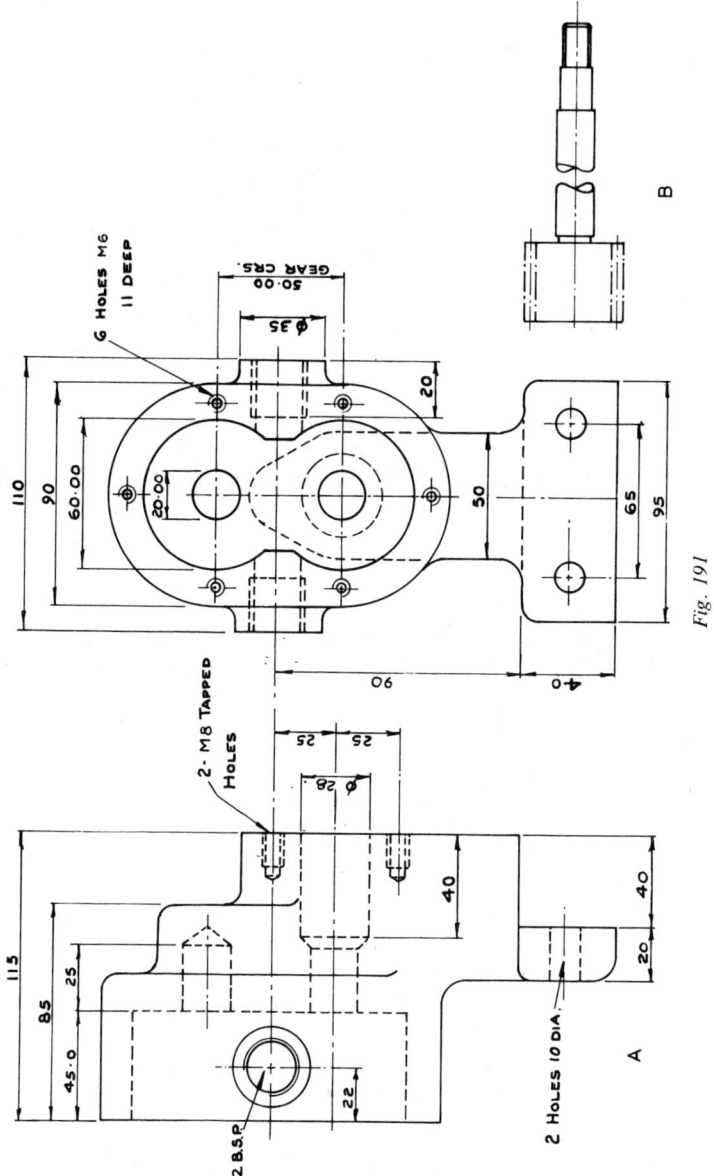

Fig. 191

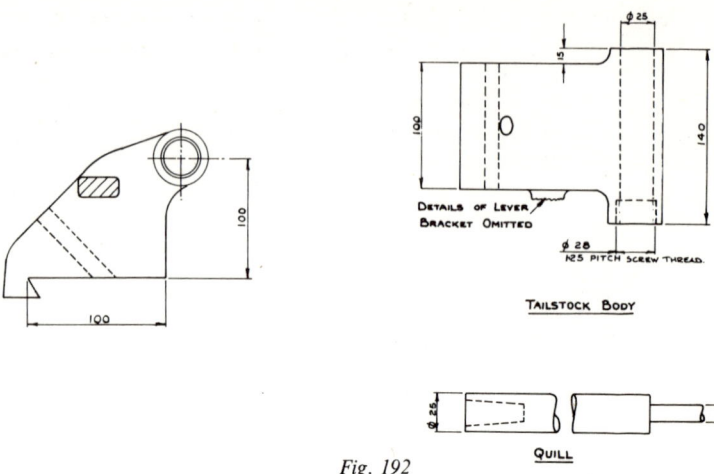

TAILSTOCK BODY

QUILL

Fig. 192

Draw an assembly of the tailstock adding any details required to complete the mechanism. The details should be ballooned and parts listed. Any dimensions not given should be estimated.

(C.G.L.I.)

18. Make detailed working drawings of all components required to complete the assembly of the spring loaded tailstock (Fig. 192 and question 17). The casting need not be detailed.

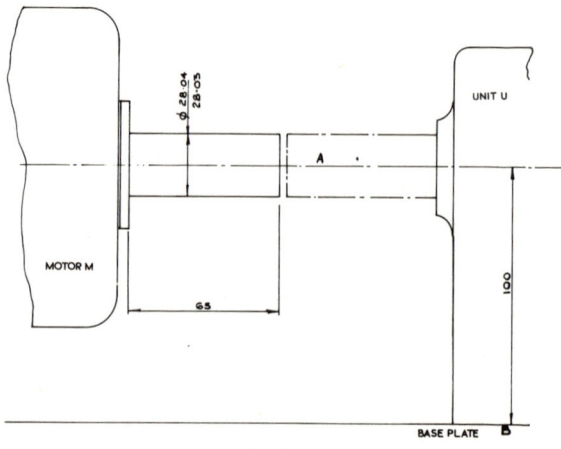

Fig. 193

19. A machine unit U has to have an input drive provided by an electric motor M as shown in Fig. 193. The unit and the motor stand on a common base plate B which can be drilled or otherwise machined if necessary. The unit input shaft can be of any reasonable length, but not more than 40 mm diameter. The motor shaft has the dimensions shown.

   Show by drawing, or a sketch in good proportion, the main features of a simple dog clutch which will complete the drive but allow it to be disconnected quickly when required.

(C.G.L.I.)

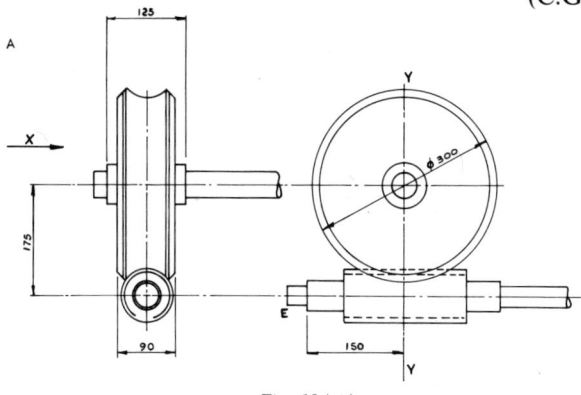

Fig. 194 A

20. Fig. 194A shows details of a small worm and wheel unit. It is intended to give the draughtsman sufficient information to enable him to work out a proposal for a casing to contain the drive. Fig. 194B shows the arrangement for a bearing at the

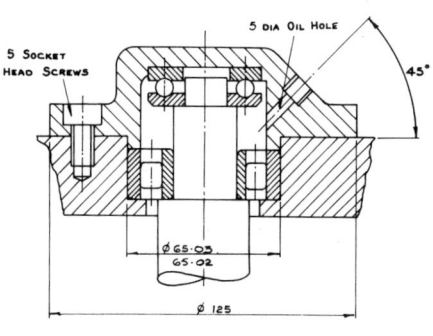

Fig. 194 B

213

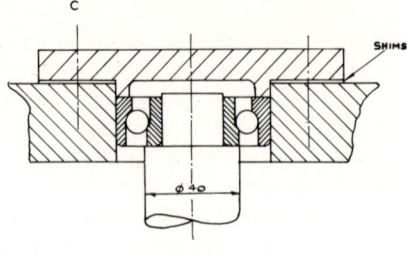

Fig. 194 C

end of the worm, and Fig. 194C that for the end of the worm-wheel.

    (i) Make a neat drawing to a suitable scale to show an outside view in direction X and a cross section on YY of any suitable form of casing with the worm-wheel assembled;

    (ii) state a suitable material and method of construction for the casing.

(C.G.L.I.)

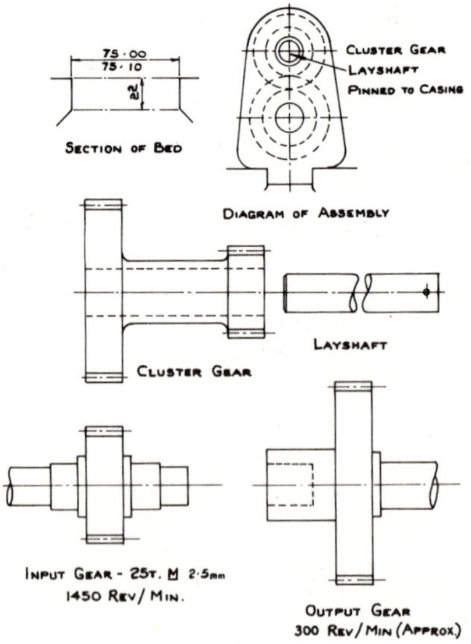

SECTION OF BED

75·00
75·10

DIAGRAM OF ASSEMBLY

CLUSTER GEAR
LAYSHAFT
PINNED TO CASING

CLUSTER GEAR

LAYSHAFT

INPUT GEAR - 25T. M 2·5mm
1450 REV/MIN.

OUTPUT GEAR
300 REV/MIN (APPROX)

Fig. 195

21. Fig. 195 shows details of the input gear, output gear and lay-shaft cluster gears for a machine tool drive. The input shaft is

driven at 1 450 rev/min and the output shaft is to run at approximately 300 rev/min. All gears are of standard involute form and proportions 2·5 mm module and 25 mm wide. The input gear has 25 teeth and the inner end of its gear shaft is extended beyond its bearing, so that the output gear shaft can run freely supported on this extension.

Determine suitable tooth numbers for the gear wheels and draw two views part sectioned as required, to show the assembled gear box complete with casing, bearings and provision for lubrication. The gear box is to sit on the bed of dimensions shown with the layshaft vertically above the powershafts as in the diagram of assembly. Any type of bearing may be selected and any dimensions not given are to be provided.

Balloon all the parts and draw up a parts and materials list.

(C.G.L.I.)

22. With reference to question 21 and Fig. 195, make detailed working drawings of *all* the component parts of the gear box, with the exception of standard items such as screws or dowels.

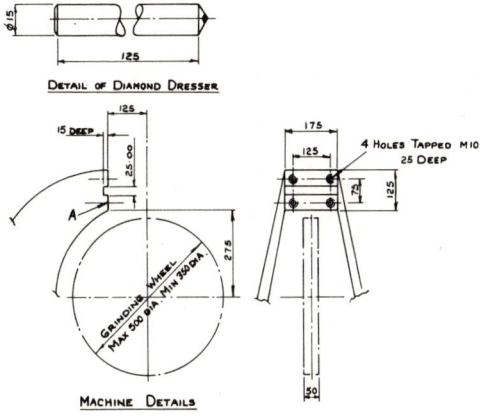

Fig. 196

23. Fig. 196 shows two elevations of a grinding wheel and part of the machine frame of a grinding machine. The overhanging piece was used to hold the guards over the wheel. The machine is being re-equipped and the guard is no longer required. It is required to mount a permanent wheel dresser on the machine.

215

The wheel dresser is to be fitted with a diamond tool as shown in the figure. Face A is already machined and has in it four tapped holes and a slot.

Design a suitable wheel dresser and show it mounted on the machine. The design should incorporate a method of feeding the diamond into the wheel and a method of traversing the diamond across the wheel.

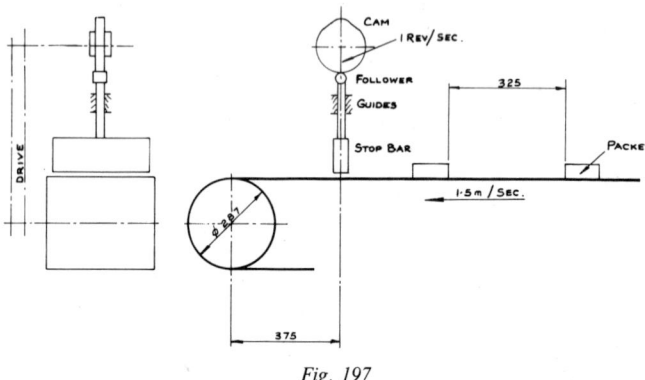

Fig. 197

24. Fig. 197 shows the outline of a collect/release mechanism used on a conveyor belt to assist packaging. The device consists of a cam-operated stop bar which descends and stops the passage of four packets. When four have been collected, the stop bar lifts and allows the four packets to continue their progress on the continuous moving belt which travels at 1·5 m/s. As soon as the packets have passed under the stop bar, the latter descends again to collect four more packets. Each packet is 50 mm wide and 25 mm thick and there are 325 mm between each packet.

Design a suitable cam-operated mechanism which fulfils these conditions. The cam revolves at 1 rev/sec and is suitably driven from the conveyor roller. Outline the type of drive envisaged and add the functional dimensions to the drawing.

The cam profile should also be drawn.

25a. Fig. 198 shows the centre line and bed dimensions of a small centre lathe.

Design a suitable tailstock for this lathe and prepare a parts list for the assembly suggesting suitable materials for the parts.

The tailstock should be provided with a method of locking

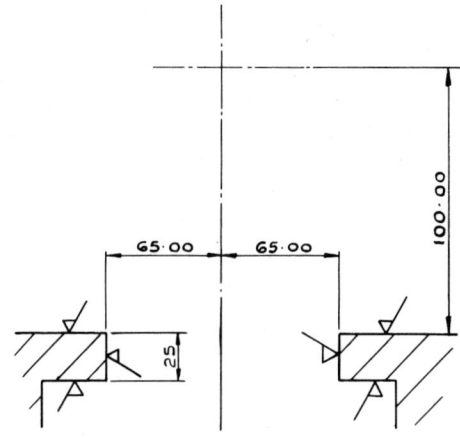

Fig. 198

the barrel in position, some means of securing the tailstock to the bed by using an integral locking lever and 13 mm adjust ment for swinging over the body of the tailstock for taper turning.

b. Make a fully dimensioned detail drawing of the tailstock body.

26. In Fig. 199 shaft B is to be driven by shaft A. The drive must be positive in one direction but should disconnect when the

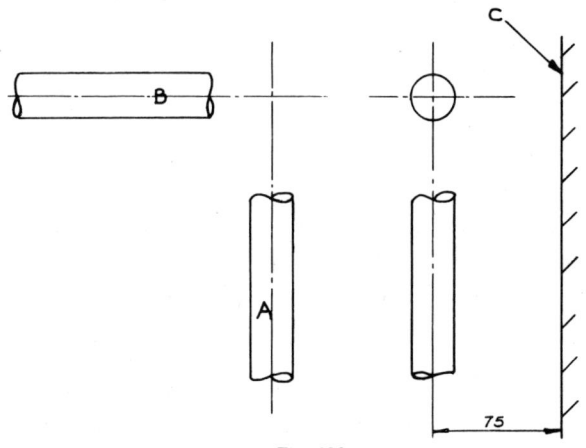

Fig. 199

217

drive is reversed. Re-engagement of the drive should be done manually. In addition, the drive must slip if overloaded and some provision must be made to vary the load at which slip will occur. The shafts are not to be enclosed, but need supporting by providing suitable bearings which can be attached to the machined surface C.

Prepare a general arrangement of the drive showing how the conditions can be met.

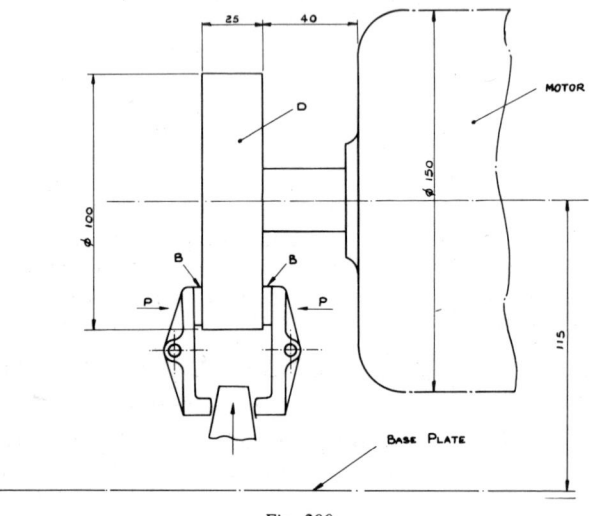

Fig. 200

27. Fig. 200 shows the rear end of a workhead which consists of an electric motor carrying a chuck which is used to hold components for burr removal and polishing. It is desired to put a brake on the workhead by using a drum D at the opposite end of the motor spindle from the chuck. A suitable arrangement would be by using brake blocks B to which pressure is applied by a wedge and simple lever system which could be operated by hand. It is estimated that approximately 400 N would be required at P.

Make a neat drawing or sketch in good proportion to show your proposals for a suitable simple brake mechanism, either along the lines suggested, or by any other means which must not be too complicated or costly. Show clearly how the brake works and make a parts list showing the materials required.

(C.G.L.I.)

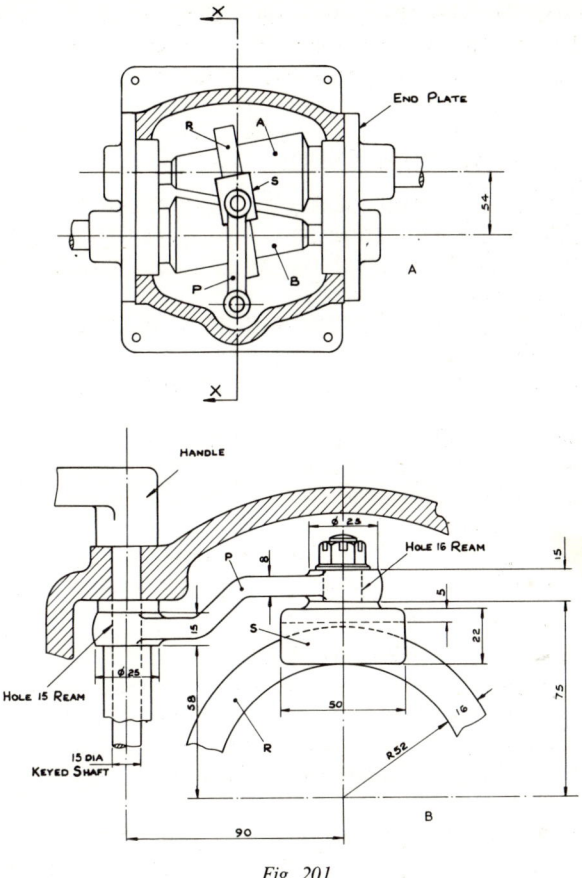

Fig. 201

28. Fig. 201A shows, in simplified form, the operating parts of a variable speed drive. Two cones of equal angle are mounted with their axes parallel and in the same plane. A rotating ring, R, encompasses the cones and is in contact with them. Cone A drives the ring, which in turn drives cone B. The ring is supported top and bottom in shoes, the form of which is shown in Fig. 201B; these allow it to rotate but, as the arm, P, swings, the ring is moved to a new position.

Using the details and dimensions given in Fig. 201A and B draw, scale full size, an end elevation which is a cross section

219

on XX. This should show the ring and its actuating mechanism and the approximate shape of the casing. The ring R may be shown circular. Dimensioning is not required.

(C.G.L.I.)

MISCELLANEOUS EXERCISES

1. Sketch or draw neatly, in good proportion and showing clearly the constructional features, any one of the following:
   - (*a*) a clamp-tip lathe tool-holder, as used for cemented carbide tips or bits;
   - (*b*) a grinding wheel collet with an arrangement for balancing;
   - (*c*) a vee-slide with adjusting gib or strip.

(C.G.L.I.)

2. Explain the importance of
   - (*a*) accuracy of alignment,
   - (*b*) load per unit area,
   - (*c*) lubricant supply,

   in making a choice of bearing metals. Give examples of common metals and their application, to illustrate your points.

(C.G.L.I.)

3. (*a*) What is meant by spheroidal cast iron?
   (*b*) Compare briefly the general properties of spheroidal graphite cast iron and of ordinary cast iron, and show how the uses to which cast iron may be put have been greatly extended by this new form of material.

(C.G.L.I.)

4. Give the approximate composition of any common zinc-base die-casting alloy. Explain the advantages and limitations of this material, and give three typical applications.

(C.G.L.I.)

5. Sketch or draw neatly the main features of a fixture suitable for holding the bracket shown in Fig. 202 whilst the slot is machined by end milling. It may be assumed that other machining, including the 20 mm diameter hole, has been completed.

   It is not necessary to dimension the fixture fully, but the four important dimensions that affect the successful working of the fixture should be shown. Name the various parts of the fixture.

(C.G.L.I.)

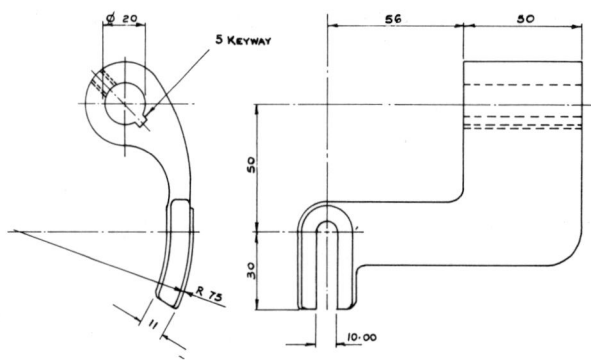

Fig. 202

6. Re-draw the cast iron lever fork shown in Fig. 202, so that the drawing contains all necessary dimensions and instructions for the manufacture of the part. The bore, the 10 mm slot and the 75 mm radius face, require machining. The drawing should satisfy the requirements of the patternmaker as well as those of the machine shop. Suitable tolerances should be shown, in particular on the slot and the bore.

(C.G.L.I.)

7. The sleeve shown in Fig. 203 is machined all over from cored cast iron stick. In use in a machine tool it slides along a shaft, acting as a rack. Main dimensions only are shown. Re-draw the sleeve in a manner suitable for a manufacturing drawing,

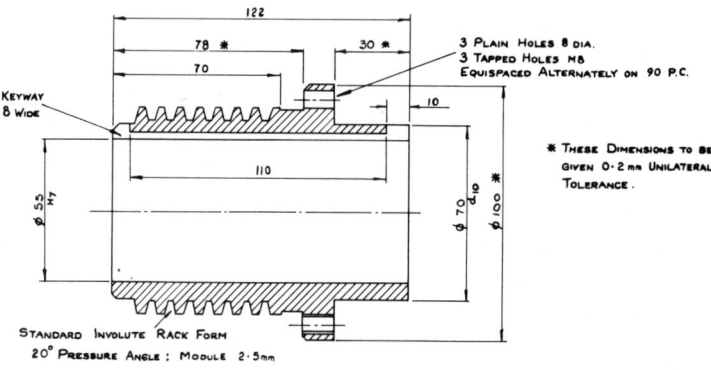

Fig. 203

giving all necessary dimensions and instructions, including suitable tolerances.

In answering this question, use may be made of standard conventional representations and of the limits and fits data sheet provided.

(C.G.L.I.)

8. The thread cut on the sleeve shown in Fig. 203 is of standard involute form. During manufacture the form of the thread will be checked by profile projection.

Draw accurately a master form for the outline of two threads using any normally available magnification, which must be stated. The master form should show a 0·1 mm tolerance at the root and crest of the thread, the purpose of which is to ensure clearance in use.

(C.G.L.I.)

9. The fork shown in Fig. 202 is designed to fit onto a shaft with a key and a grub screw so that it will transmit a torque and be restrained from moving endways.

Sketch or draw three alternative methods of fixing to achieve the same effect and mention briefly any advantages or disadvantages which each has, compared with the existing method. Alterations to the form of the casting may be suggested, if desired.

(C.G.L.I.)

10. (a) Explain briefly what is meant by the term *tolerance zone* as used in positional or geometrical tolerancing.

(b) Show the application of positional or geometrical tolerance applied to two dimensions taken from Fig. 202 or 203 and briefly explain the effect of each in relation to manufacture.

(C.G.L.I.)

11. (a) What is meant by *bearing pressure* on a journal bearing? State briefly how this influences the choice of bearing metal.

(b) Give an example of a low melting point bearing alloy, quoting its approximate composition, main characteristics, and typical applications.

12. Give an example of a material in the group known as *high tensile bronzes*, quoting its approximate composition, main properties and a typical application.

(C.G.L.I.)

13. Some materials are used in engineering construction because of their special properties in service, although these are accompanied by considerable manufacturing difficulties.

    Give an example of such a material with brief details of its character and properties, and state how the manufacturing difficulty is usually overcome.
    (C.G.L.I.)

14. Explain, with the aid of sketches to show constructional details, any two methods of adjusting a bearing for wear or play. The action necessary to make the adjustment should be quite clear.
    (C.G.L.I.)

15. (a) Explain briefly the difference between malleable and spheroidal graphite cast irons.
    (b) Choose an application for which either material might be suitable and outline the factors which would be considered in making a decision which to use.
    (C.G.L.I.)

16. (a) Outline the main advantages of sintered materials for use in cutting tools.
    (b) Name two sintered materials commonly used for cutting tools and indicate, with reasons, the type of application for which they are suitable.
    (C.G.L.I.)

17. (a) Give a typical application of a magnesium-base alloy and outline the main properties of this material.
    (b) Explain one peculiarity of magnesium-base alloys with respect to
        (i) machining, and
            either
        (ii) directionality
            or
        (iii) electro-chemical behaviour.
    (C.G.L.I.)

18. Fig. 204 shows a swivel angle-plate casting and gives some information about the finished plate.

    Make a drawing suitable for issue to the machine shop giving all dimensions and instructions necessary for the angle-plate to be machined.

    Diameter tolerances can be obtained from B.S. Data Sheet No. 1 provided. Geometrical tolerances need not be shown.
    (C.G.L.I.)

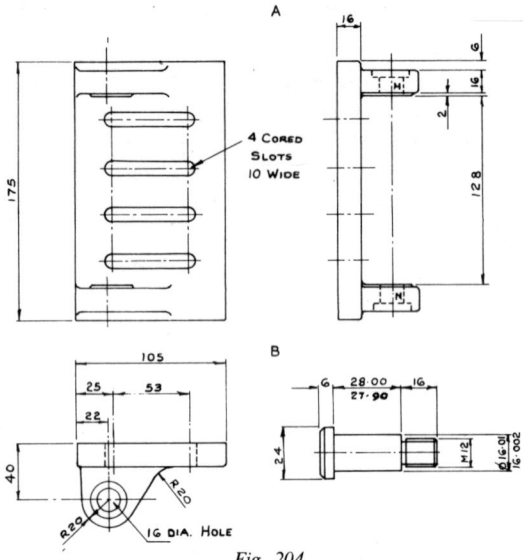

Fig. 204

19. Fig. 204 shows a swivel angle-plate which, when fully machined, has two holes M and N which are counterbored to suit the pin shown in Fig. 204(A).

    (*a*) Sketch any form of tool using an inserted tool-bit, suitable for producing the counterbore. Dimension the working end of the tool.

    (*b*) Specify clearly the nature of the tool-bit, showing by a sketch the cutting angles, and giving the type of material or materials used.

                                               (C.G.L.I.)

20. In the swivel angle-plate shown in Fig. 204, the two holes M and N must be accurately co-axial with each other, and their common axis must be parallel with the face of the plate.

    (*a*) Show how these requirements can be indicated on the drawing by using geometrical tolerances. Briefly explain the effect of the tolerances.

    (*b*) The 16 mm diameter hole must be finished by honing to a quality of 1·3 microns. Show how this would be indicated on the drawing.

                                               (C.G.L.I.)

21. Using simple line diagrams, show the linkage system of movement of any mechanical comparator. Explain briefly how friction is minimized and what magnification is obtained.

                                               (C.G.L.I.)

22. Using line diagrams, show the main details of any method of traversing a machine slide by screw and dial with vernier. Suggest a dial size and vernier graduation to give increments of movement of 0·02 mm in assuming a screw of 5 mm pitch.

(C.G.L.I.)

23. Give the approximate composition of any commonly used metal in the group known as nickel-base alloys, outlining its main properties and giving typical applications.

(C.G.L.I.)

24. Suggest a material suitable for each of the following applications. Give details of the properties which make it suitable for the application, the form in which it would be supplied and any treatment which it would need before use.
    (*a*) A plain bearing for an accurate high-speed spindle.
    (*b*) A plug gauge for continuous use in abrasive conditions.
    (*c*) Non-ferrous gears for food processing machinery.

(C.G.L.I.)

25. The sub-assembly shown in Fig. 201B consists of an arm P carrying a shoe S which fits on and guides the ring R of width 16 mm. The shoe is attached to the arm by a pivot which is integral with the shoe. The arm is supplied as a forging.

    Make a production drawing of the shoe and arm suitable for issue to the machine shop and giving the dimensions and other information required for machining. Use the Data Sheet, B.S. 4500 'Primary Selection of Fits', to obtain the tolerances.

(C.G.L.I.)

26. Fig. 205 shows a form to be machined on a capstan lathe using a tangential type form tool. If the tool is to have twelve degrees top rake and six degrees front clearance:
    (*a*) draw the true cross section of the tool and the dimensions necessary for making it;
    (*b*) sketch a typical form tool-holder for a form tool of this type;
    (*c*) specify a suitable material for the tool if it is to operate on 0·3 % carbon steel.

(C.G.L.I.)

27. (*a*) Design a 'scrap layout' (or blanking layout) for the component shown in Fig. 206 making the most economical use of the strip and bearing in mind the finished form.

    Minimum scrap width is to be $2\frac{1}{2}$ times the strip thickness.

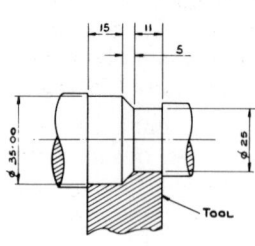

Fig. 205

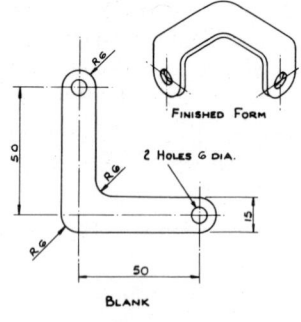

Fig. 206

(*b*) Dimension
   (i) the feed length, and
   (ii) the strip width calculated to the nearest commercial size.
(*c*) Make drawings or sketches to show the punch and die layout and dimension the die apertures.

(C.G.L.I.)

28. (*a*) Explain what is meant by 'accumulation of tolerances', e.g. in an assembly.
   (*b*) Explain briefly the advantages of positional tolerancing for hole centres as compared with centre distance limits.
   (*c*) Show how any one of the following conditions can be controlled by the application of a geometrical tolerance to a typical component and explain its effects:
   (i) parallelism   (ii) symmetry   (iii) squareness.

(C.G.L.I.)

29. Common reasons for the use of sintered metal parts are:
   (*a*) properties which are unobtainable in normal alloys;
   (*b*) parts which are non-producible by more conventional means;
   (*c*) economy of manufacture.
   Give for each of the above reasons an example of a typical component and for each example justify your choice in relation to its application.

(C.G.L.I.)

30. (*a*) Explain what is meant by 'directional properties' in

relation to its forms of supply such as sheet and strip. Why are such properties important in manufacture?

(*b*) List the main physical properties normally quoted in material specifications for cast iron and explain how they differ for sand cast malleable and spheroidal graphite (s.g.) cast irons.

(C.G.L.I.)

31. (*a*) Give the approximate composition of a typical zinc-base die-casting alloy and outline its main physical properties as cast.

(*b*) Die-casting of zinc-alloys by the hot-chamber method is fast and cheap and gives a product of good appearance requiring little or no finishing. Outline the main limitations in the application of such die-castings.

(C.G.L.I.)

32. The arm P shown in Fig. 201B is supplied as a steel forging. It has to have two holes drilled in it for reaming as a second operation after the ends of the arm have been faced by straddle milling.

Draw or sketch, in good proportion, a drill jig showing how the arm is located and clamped, and the holes are drilled.

(C.G.L.I.)

33. Discuss the tendency to replace phosphor-bronze with nylon as a bearing material. Outline the advantages and disadvantages of nylon in its use as a bearing material. State any particular application where it would be inadvisable to use nylon.

34. Select a component which could be manufactured equally well by fabrication, forging and casting. Discuss the factors involved in the selection of each manufacturing method for this component.

Index

© Cassell & Co. Ltd 1968